Ernest Rutherford and the Birth of Modern Physics

ERNEST RUTHERFORD and the Birth of MODERN PHYSICS

MATTHEW WRIGHT

SCRIBE

Melbourne | London | Minneapolis

Scribe Publications
18–20 Edward St, Brunswick, Victoria 3056, Australia
2 John St, Clerkenwell, London, WC1N 2ES, United Kingdom
3754 Pleasant Ave, Suite 223w, Minneapolis, Minnesota 55409, USA

Published by Scribe 2025
Published in New Zealand by Oratia Books 2025

Internal pages designed by Sarah Elworthy

Printed and bound in the UK by CPI Group (UK) Ltd, Croydon CR0 4YY

Scribe is committed to the sustainable use of natural resources and the use of paper products made responsibly from those resources.

978 1 761381 17 1 (Australian edition)
978 1 915590 96 1 (UK edition)
978 1 964992 16 7 (US edition)
978 1 761386 47 3 (ebook)

Catalogue records for this book are available from the National Library of Australia and the British Library.

scribepublications.com.au
scribepublications.co.uk
scribepublications.com

Contents

Introduction

How lucky we are to follow Rutherford, and how lucky we are to have a chance to devote a lifetime to it.

— D.R. Inglis, Rutherford Jubilee International Conference, September 1961[1]

THIS IS A BOOK ABOUT ERNEST RUTHERFORD and his place in the birth of modern physics. The science that Rutherford and his colleagues discovered during a few heady decades from the 1890s was stranger, cooler and more amazing than anyone had ever imagined. In a succession of bold strokes these pioneering physicists revealed a startling vision of largely empty atoms, elements that spontaneously disintegrated to release unimaginable energies, and hurtling particles and atomic forces of colossal power. All floated in a dynamic reality where mass and energy were aspects of the same thing, where the very fabric of space and time could twist around on itself, and where particles behaved in ways that seemingly defied comprehension.

This was not magic: it was science, and the key figures — Ernest

Rutherford, Marie Curie, Albert Einstein and Niels Bohr, among others, became household names, and few more so than Rutherford. When he died in October 1937 the *New York Times* mourned the loss of a man who had achieved 'immortality' and 'Olympian rank' in his lifetime.[2] This was not hyperbole. In just a few dramatic years from 1898, Rutherford defined the underlying principles of radioactive decay, identified what 'radiation' actually is, explored the particles and forces involved, and went on to define the concept of the atomic nucleus that lay behind these phenomena — discoveries that reset the basis of physics and chemistry.

These achievements alone would have given Rutherford headline space in the history of science. But these were merely the beginning of his restless investigation into the unknown. His student Niels Bohr infused Rutherford's concept of an atomic nucleus with early principles of quantum mechanics, establishing the basis of an atomic model that remains current today. Then, during the First World War, Rutherford did what was widely regarded as impossible: he split atoms. In the process — though he was too careful a scientist to admit it too soon — he transmuted one element into another, achieving the goal that had eluded alchemists, and which had been rubbished by the sober scientists of Western civilisation as magic.

Later, Rutherford was behind the British team that split atomic nuclei on a shoestring in 1932, beating a well-backed American team. In and around his other work Rutherford experimentally verified the accuracy of Max Planck's equations, confirming the existence of quantum mechanics.[3] He defined Avogadro's number — an elusive holy grail of chemistry that has application today in everything from research to industry. He provided the first realistic estimates of the age of the Earth, helped pioneer radiotherapy as a treatment for cancer — including establishing the Cambridge Diploma in Medical Radiology and Electrology[4] —took a major role in the invention of sonar, and helped develop the basic technology for detecting ionising radiation, as well as ionising smoke detectors and ultrasound machines.

In short, Rutherford's work stands behind numerous major

branches of modern science and underpins wide swathes of modern life, industry and technology. His colleagues were well aware of his achievements. Albert Einstein, apparently in awe, referred to Rutherford as a 'second Newton'.[5] James Jeans, himself a leading figure of his day, considered Rutherford one of the greatest scientists that ever lived.[6] Some of these comments came soon after Rutherford's death and have to be seen in that context — but the sentiment never faded. In 1989 the physicist Samuel Devons declared that much of physics and chemistry today had been profoundly influenced by what Rutherford discovered. Indeed, the whole of modern science, Devons felt, was itself the sequel to Rutherford's work.[7] This last was hyperbole, but not by much.

Part of the reason for Rutherford's status was that his hands-on work was only part of his contribution, and his colleagues knew it. At the turn of the twentieth century Rutherford worked at McGill University in Canada with such verve he was credited with establishing physics as a viable study in that country.[8] His subsequent roles heading the physics department at Manchester University, and later running the prestigious Cavendish Laboratory at Cambridge University, put him in a position to direct teams he had picked, all working at the cutting edge of the field. He had a knack for finding talented individuals, then pushing them down lines that became crucial to the new science; or which — in the case of Ernest Marsden — led to physics being the hot new field taught at major universities.[9]

Rutherford's rise to international stardom opened doors in his original home. In the early twentieth century New Zealand was wallowing in 'cultural cringe', the idea that status could be earned only by achievements overseas, ideally in Britain. Rutherford was arguably the greatest of all. By the early twenty-first century, when popular memory of most of New Zealand's twentieth-century heroes had faded, Rutherford was one of a handful who remained a household name in his home country, commemorated in roads, buildings, awards and on the highest-denomination New Zealand banknote, the $100. I was on the committee that selected him for that note, and the decision was made to place him there because, of all the figures on the shortlist, Rutherford was the most accomplished.[10] The portrait, a carte-de-visite by a Manchester photographic studio, revealed Rutherford in his prime, a young man driven to discover the secrets of the universe.[11]

This book explores Rutherford's work and places it in context of the birth of modern physics, the bold vision whose basis emerged to science over roughly four decades from the 1890s. These concepts remain the basis of physics today. The story begins by establishing place: a brief explanation of how Western science emerged and where the new physics came from, showing how the 'classical' ideas of Newton and his successors initially shaped Rutherford's approach. Much of the science history outlined here is specific to the questions Rutherford answered and the ideas he brought together. From there the story follows Rutherford's journey through the world of electrodynamics, the hot new topic of the late nineteenth century. We explore how that led him to discover the deepest secrets of the atom, and how in the process he overthrew older ideas.

To give this place and show how Rutherford's work interlinked with the other discoveries of the new physics, the book wraps this story with an outline of the work by his colleagues as they explored the new territories being revealed to them. We explore this in sequence, just as the great minds of the early twentieth century did. Some of their explanations are no longer current today: they were on a journey, and this book traces that path. I have made clear where ideas have been superseded. Current explanations are given in sidebars or as comments where necessary. And along the way we will learn just how human that story was. For all the conceits of science as an objective pursuit, those involved were real people who lived real lives. The frameworks of their societies, their life experiences and social requirements of their day were part of the mix. Rutherford was no exception.

One point needs to be made regarding terminology. I have used the term 'theory' in its scientific sense. In ordinary English, theory means an unproven conjecture. But this is the reverse of what it means in science, where an unproven conjecture is a 'hypothesis', and 'theory' means a concept that has been proven true by empirical data such as experiments and tests that could falsify the proposal. Why are such concepts called 'theories'? One reason is that Western science demands revision on the basis of new data. Sometimes a theory is even thrown out altogether and superseded by a dramatic new suite of ideas. The process by which major shifts of theory happen in Western science — increasing unease with one theory, a crisis of thought, and then a new paradigm — was examined by Thomas Kuhn (1922–96), among others. Modern physics itself is a product of such a revolution.

The book is sourced from a range of papers and archival sources. Rutherford was not a great hoarder: when asked to write a memoir of his early days in Cambridge he admitted he had not kept any letters and realised that his memory was not as detailed as he liked.[12] A remarkable quantity of written material nonetheless survives. Archival sources used for this book include the extensive Rutherford collection in the Alexander Turnbull Library (ATL) in Wellington, New Zealand. This contains copies of Rutherford's outward letters, correspondence from his family, numerous papers, biographical notes and personal recollections that his student Ernest Marsden obtained and assembled for Rutherford's first official biography, and for a second biography Marsden intended to write himself. Supplementary material held at ATL includes microfilm copies of British records, such as the extensive Rutherford-Bragg correspondence. The other main archival source is the Rutherford collection at Cambridge University Library (CUL) in the UK, which includes Rutherford's Cavendish material, his surviving notebooks — some still radioactive — and many hundreds of pages of notes and inwards correspondence. These give a remarkable insight into the man and allow a number of persistent myths about Rutherford and his work to be given place.

This book is co-published by Oratia Books and Scribe, and I am grateful for the support of my publishers, Peter Dowling and Oratia Books in Auckland; and Henry Rosenbloom and the team at

Scribe in Melbourne. Mike Bradstock provided solid editorial advice. I am particularly grateful to Lemuel Lyes, who provided professional research services in the UK. It is thanks to his work with the Rutherford papers that I have been able to use such a complete record of the Cambridge material. Dr John Campbell of the University of Canterbury read the New Zealand biographical chapters and I am grateful for his comments on the drafts. I also thank the staff of the reading rooms at CUL and of the ATL for their support, along with Murielle Baker of Rocket Lab for her go-ahead to use pictures of the New Zealand-designed rocket propelled engines bearing Rutherford's name. My thanks also go to Professor Mary Fowler, Rutherford's great-granddaughter, for her permission to quote from Rutherford's unpublished documents held by CUL.

Matthew Wright
June 2025

1 | The summer of '96

IN THE ENGLISH SUMMER OF 1896 a professor and his student set up a curious apparatus in the Cavendish Laboratory on Free School Lane, Cambridge. The laboratory was part of Cambridge University, and one of a handful of specialised physics institutions of the day and arguably the most prominent at the time.[1] The professor, Joseph John Thomson — 'J.J.' to his friends — was a small man, just 40 years of age with a straggling moustache and hair kept over-long for the day.[2] His student, Ernest Rutherford, was 25, tall and broad, with a fashionably full moustache and booming voice, and hailed from New Zealand. He was one of the first non-Cambridge students to take their post-graduate programme and had already made a name for himself with radio wave detection equipment.

The equipment they were assembling was hand-made for its purpose, as most laboratory equipment of the day usually was. It was built around what Rutherford called a 'bulb',[3] usually known as a Crookes tube after its inventor, the physicist William Crookes (1832–1919).[4] It looked like an elongated light bulb: a pear-shaped glass container with wiring inside. The interior was almost entirely a vacuum, the air inside laboriously evacuated with a hand-worked mercury pump. This did not wholly remove all the air, but the trace that remained was crucial to Thomson and Rutherford's work. This particular tube had been made for them by Thomson's assistant Ebenezer Everett, a talented glass-blower who also forbade Thomson from handling it. Thomson, it seemed, was known for his clumsiness.[5]

The tube contained two bare-metal plates. The one at the narrow end was known as a cathode — technically, a 'cold cathode' because it was unheated. The other plate, part way along its length, was called an anode. These were connected to a Ruhmkorff induction coil, a device that could supply high-voltage electricity. When switched on, electrical current flowed from the cathode through the residual air in the tube to the anode — and as Thomson put it, that caused something to be 'shot off' from the cathode, past the anode and through the glass at the flattened end of the tube,[6] which lit up with an eerie green glow.[7] This effect had been observed as early as 1859 when the first cold-cathode tubes were devised, and in 1876 the 'something' that caused it was given the name *Kathodenstrahlen* (cathode rays) by the German physicist Eugen Goldstein (1850–1930). Nobody agreed on what they were: Crookes himself thought these rays were simply fast-moving atoms, but the German physicist Heinrich Hertz (1857–94) thought they were electromagnetic waves.

To that extent there was nothing particularly new about the Crookes tube Thomson and Rutherford were setting up that summer's day in 1896. But what Thomson was primarily interested in was a discovery made the previous December by the German physicist Wilhelm Röntgen (1845–1923), who found that rays pouring out of the end of these tubes could penetrate solid matter and produce ghostly images on a cardboard screen painted with barium platinocyanide. He dubbed the emissions 'X-rays' to reflect their unknown nature. The name stuck with the public, though the emissions were also called 'Röntgen' rays by his colleagues.[8]

Thomson thought he might be able to use these rays as a tool to extend his research into the relationship between electricity and matter. This field was the hot new area of investigation of the day, one of several areas to which physicists were drawn on the back of broader investigations into a newly discovered phenomenon, electrodynamics — the science of moving electromagnetic fields. The basic ideas built on the work of the prominent early Victorian-age experimentalist Michael Faraday (1791–1867) and the mathematician J. Clerk Maxwell, but by the 1890s were being taken in new directions. Thomson's problem was that he needed to measure infinitesimal electrical values. Rutherford had recently arrived in Cambridge on the strength of post-graduate work in New Zealand which revolved around doing just that, and was shoulder-tapped by Thomson. Their first step involved replicating Röntgen's rig. Rutherford was fascinated by the gadget, later writing to his mother, 'You know one can see the bones of the hand and arm, and coins inside with the naked eye.'[9]

Thomson and Rutherford worked with the mystery rays for weeks, blasting various gases — including chlorine, hydrogen, mercury vapour and ordinary air — and measuring the way in which these gases conducted electricity.[10] It was exacting work. Rutherford typically arrived in the laboratory at 10 a.m., and apart from a somewhat lengthy break for lunch — where he swapped out with Thomson — worked through to 6.30 p.m. Thomson joined him typically from 4 or 5 p.m.[11] The fact that Thomson was forbidden to handle the Crookes tube that lay at the heart of the apparatus didn't stop the devices burning through the cathode — which dropped the rate of X-ray output — and Everett was kept busy making new ones.[12]

The spectre of these two scientists casually exposing themselves and their assistant to ionising radiation seems extraordinary today.[13] But in that summer of 1896 the nature of X-rays was unknown. This had not stopped the world going X-ray mad.[14] Röntgen's discovery was an overnight sensation, which in hindsight wasn't surprising. The idea of looking through solid objects was akin to magic. Yet nineteenth-century science had made it possible, and for a while these mystery rays became the latest craze. Fluoroscopes — essentially the apparatus Thomson and Rutherford set up — were relatively easy to make, and public X-ray machines became popular in amusement parks, circuses and other places around the Western world, purely so that people could gaze at their own hand bones.

The idea even leaked into fiction. That year the social commentator and science enthusiast H. G. Wells was penning a new novel, riffing on recent events in Africa where a handful of British soldiers used machine guns to mow down 1500 Ndebele warriors. Wells wondered what might happen if the British capital were attacked by an enemy with weapons superior to their own. By 1896 he had his story of Imperial nemesis under way. And in this age of rays at the cutting edge of science, his invading Martians inevitably ended up with one as a weapon.[15]

Scientists and engineers took a more sober view, but the new field was exciting in other ways. Until the late nineteenth century Western science pivoted around a basic set of principles broadly devised by Sir Isaac Newton (1642–1726) and his colleagues in the late seventeenth century, concepts in turn framed by the geometry devised by the Greek philosopher Euclid nearly 2000 years earlier. The underlying ideas were evident to anybody: drop something and it fell. Push an object and it would respond by moving. Straight lines were obviously straight, curves were measurable, fixed distances were obviously fixed, and so it went on. The world was, in short, exactly as it seemed to everyday human perception — and Newton, along with his friends Robert Hooke, Edmond Halley and a few others, wrapped a good deal of mathematics around it to prove the point.

This science — 'classical' physics — was part of a remarkable evolution in Western thinking from the late seventeenth century, which leaked into society and ultimately helped drive the Industrial Revolution. Newton's equations for gravitation even made it possible to discover a new planet, Neptune. But by the late nineteenth century new developments were changing this concept. Western science led the world, and in the last decades of the century electricity became the new wonder-fluid of the age. From a public perspective science was on the ascendant, giving Western civilisation a string of triumphs that transformed everyday life.

What the new technologies did not reveal, though, were the growing cracks spreading through classical science as a result. The problem was simple enough: the electromagnetic forces that gave the world its new age of wondrous electrical devices didn't work the way classical physics said they should. The crisis took time to build. The relationship between magnets and electricity was made clear in the early nineteenth century: they were both aspects of the same phenomenon, electromagnetism. So was visible light. The whole field became known as electrodynamics and led to a raft of mysteries that remained unsolved in the 1890s. Even the way that electromagnetic energy was transmitted challenged thinking. Visible light had long been considered to be a series of waves or ripples in an invisible and undetectable 'luminiferous aether'. Logically, the rest of the electromagnetic spectrum as it emerged to physicists during the last decades of the nineteenth century, was transmitted the same way. But there was a problem: repeated experiments to find or measure this aether during the late nineteenth century were unsuccessful.

Then there was the problem of 'cathode rays'. Sir William Crookes, inventor of the vacuum tube that produced them, thought that these rays were streams of particles. His colleague Arthur Schuster, professor of physics at Victoria University of Manchester, concurred. The only question was: what *sort* of particles? By this time chemists had shown that matter was made up of many different kinds of atoms, each type forming a different element. By that summer of 1896 some 76 elements had been found, most recently helium.[16] All were seen as immutable, because the atoms that comprised them were thought to be unchanging. Chemists also understood that atoms could combine to form molecules. So from the British perspective of the day, the question was whether cathode rays were atoms or molecules.

That idea wasn't shared on the other side of the North Sea, where Hertz was unable to replicate Crookes' results. He instead decided that cathode rays were a kind of light, in other words that they were waves in an invisible luminiferous aether. The fact that nobody could find any sign of the aether itself didn't matter: the rays were demonstrably waves, therefore the aether existed. Waves or particles? Aether or ... what? This wasn't the only challenge then confronting physicists as they struggled to understand electrodynamics. A lot of what they were now observing didn't seem to fit the comfortable picture Newton and his friends had built up long ago during their lengthy discussions in London coffee shops. They needed more data. And that was why Thomson and Rutherford were so busy during the summer of 1896 — as, indeed, were physicists across the continent, mostly with their own teams of colleagues and students in support.

Thomson continued to work on the issue through the year and, thanks to Rutherford's methods for electrically measuring the results was able to discover the charge-to-mass ratio, revealing that the 'rays' were actually negatively-charged particles far smaller than atoms. This was a shattering discovery. Until then, atoms had been considered the smallest possible object, unbreakable and solid — visualised, as Rutherford later put it, 'something like a billiard ball or a boy's marble.'[17] Now Thomson had found something smaller — something that clearly lay within the atom. His paper on his incredible discovery was published in 1897 and rocked the physics community. Not everybody accepted it, but Thomson went on to explore just what these particles were, setting his students to work on experiments designed to provide further data. Rutherford played due part in that process, wrote papers on his own work in this field, jointly authored others with Thomson — and with that he was on his way, a rising star in a new area of scientific endeavour.[18]

Opinion was divided as to what to call these new subatomic particles. Thomson referred to them as 'corpuscles.' Everybody else referred to them as 'electrons',[19] a name invented in 1891 by George Johnstone Stoney (1826–1911) for the smallest possible electric charge, which had not yet been discovered at that time. Thomson's 'corpuscle' carried that charge, and Stoney's nephew George FitzGerald (1851–1901) — himself a leading physicist — championed the name.[20]

In many respects the field of subatomic particle physics was born from Thomson and Rutherford's work in 1896–97. Of course all science builds on what has gone before, and this was no exception. Their work was part of broader research being undertaken in scientific institutions

THOMSON'S EXPERIMENTAL APPARATUS

This diagram shows the apparatus Thomson used to measure the deflection of cathode rays by an electromagnet. From this he was able to deduce their nature and discover the electron.

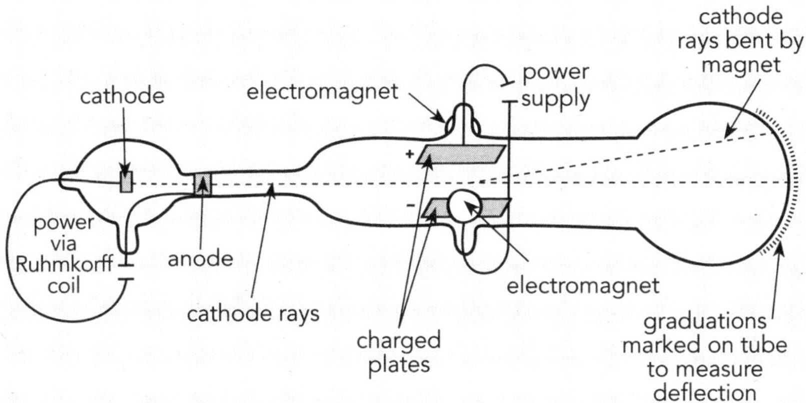

and laboratories across the world at the time. But the discovery of the first subatomic particle was a fundamental breakthrough. Suddenly a whole new field was revealed — one that became central to the way the new physics was then explored. Indeed, particle physics remains central to physics today. The closest physicists have come to a 'theory of everything' is the Standard Model of particles, because it defines everything in the visible universe, including the forces that operate it.

Both Rutherford and Thomson realised a new vista was unfolding before them. As we shall see through this book, the development of modern physics followed the effort to understand the world in terms of electrodynamics, answering the questions that arose when things didn't entirely add up in classical terms. This led to a complete revolution, including Einstein's relativity and the discovery of quantum mechanics, the principle on which electrodynamics was found to rest. The revolution lasted into the late 1920s, by which time the main foundations of modern physics were in place. Much remained to be learned, but the primary idea-set — the basic concepts of particle physics and its two 'operating systems', quantum mechanics and relativity — were there. What followed was structured around these ideas.

Rutherford played a key part in that revolution, inadvertently transforming himself into a household name. The science of his era, with its audacious explanations that were both counterintuitive and yet compelling, captured the social mood of the time. The physics was tailor-made to engage with the prevailing Western mindset of 'progress'. The major figures of the day — Rutherford, Pierre and Marie Curie, Albert Einstein, Erwin Schrödinger and Werner Heisenberg among them — became household names. These physicists and their colleagues set the foundations for all the major developments in physics that followed.

Modern physics has since shaped everyday life to the point where many of its effects are invisible. The public X-ray fervour of the late 1890s was followed by far safer uses for X-ray technology in diagnostic medicine. The intrusion of modern physics into everyday homes continued with radio — then called 'wireless' — which swept the developed world in the years after the First World War and offered new ways by which people could engage with the world. This brought dramatic change to daily life. Mid-century evenings revolved around listening to radio broadcasts, changing the way families interacted. That then evolved into television — transmitting moving pictures on FM radio frequencies.

Widespread adoption of television was interrupted by the Second World War, but this technology became the hot property of the mid-twentieth century. Once again modern physics was transforming everyday life. The glass picture tubes on which TV relied until the early twenty-first century were directly descended from the laboratory apparatus used by Rutherford and Thomson in that summer of 1896. This meant that a large slice of humanity spent their leisure hours staring directly down the barrel of an electron gun,[21] seldom realising that this technology was identical in principle to the Crookes tube. Today, radio-frequency transmission remains central to every home in the developed world, primarily through Wi-Fi and cell phone technologies, both transmitting digital data via low-power radio transmission.

Many of the terms and ideas of physics became embedded into mainstream pop culture, to the point where one of the most popular sitcoms of the 2010s revolved around the everyday life of physics geeks.[22]

The nuclear science that began in Rutherford's laboratory also changed global politics. Pop culture assigns many Second World War technological developments to the Axis powers, but in fact it was the Allies who developed most of the 'super science'. The cavity magnetron, a British invention that transformed radar, today sits at the heart of every modern household microwave oven. The atomic bomb, developed by a massive Allied programme that began in mid-1942, became an existential threat that altered the international political system, creating a spectre of imminent Armageddon that remains true today.

Rutherford directly established the foundations of that particular path and, perhaps inadvertently, inspired its trajectory. In 1933 he gave a lecture that provoked his Hungarian colleague Leó Szilárd (1898–1964) to start thinking about nuclear chain reactions, in which the detritus kicked out by one split atomic nucleus would split others and so on, exponentially increasing the energy release. By 1939, work along these lines in the United States provoked Szilárd to ask Einstein to sign a letter warning President Roosevelt of the terrible consequences if leading German physicists — all of whom Einstein knew personally — were able to build an atomic bomb.[23]

In all these ways — and more — modern physics has not merely given answers to fundamental questions about how the universe works. It has also shaped the modern world, its politics, its popular culture, and its everyday life. And that poses questions. Why did modern physics emerge when it did, in Western culture at the height of its Imperial power? What part did Rutherford play? To understand that we have to first ask not *what* Rutherford and his colleagues discovered — for the answer is a straightforward narrative, and one we will come to — but *why* these discoveries emerged in that particular time and place.

2 | The origins of Rutherford's science

A well constructed theory is in some respects undoubtedly an artistic production.

— Ernest Rutherford, 30 April 1932[1]

IN 1931 THE PHYSICISTS Otto Hahn and Lise Meitner declared that to present all of Ernest Rutherford's 'brilliant discoveries and original creations' in detail demanded first describing 'the whole development of physics and modern chemistry'.[2] They were broadly right, and to fully understand what modern physics is, Rutherford's place in its evolution, and why the main components came together in the four decades or so from the 1890s, we have to look at where this science came from and some of the ideas behind it.

At its broadest level, modern physics is the latest incarnation of a long-standing human endeavour: the effort to understand the world around us. It is also a product of the Western diaspora, derived from a set of ideas and approaches to science that emerged across Britain and Europe from the late sixteenth century. Europe's expansion across the globe during the same period brought the same idea-set to its colonies. This thinking and the philosophies that went with it — notably the concept of objectivity — shaped the nature of Western science and Western society in general.[3]

During the mid-to-late twentieth century there was a rare debate over the nature of Western science as an investigative endeavour — the 'philosophy' of the field. The usual view from within the sciences itself was founded in orthodoxy: science explored and described an objective reality, independent of humanity, framed by empirical data obtained by methods that were untainted by human influence. This was in line with the vision that originated in the West's general philosophical and scientific renaissance of the sixteenth century. Science, this view declared, was independent of personal bias or subjectivity.

An alternative view was proposed by the so-called 'externalists'. This approach harnessed the concepts of critical analysis that were developed during the mid-to-late twentieth century as tools for the humanities. By this thinking, science might well have objective outcomes, but the *path* to that understanding — including the methodologies — was always going to be shaped by the societies of its day, by the individuals involved, even by the internal rivalries and what could be called the politics of any scientific field.[4] This idea provoked intense debate from the 1980s, not least because structured objectivity — ensuring that discoveries and conclusions were untainted by emotional, cognitive, and methodological bias — remained one of the foundational pillars of the scientific approach. There was considerable pride and ego involved in applying it.

The idea that the process of scientific discovery was always going to be shaped by society and times was further explored during the early twenty-first century.[5] The general conclusion, as Philippe Stamenkovic (b. 1998) put it, was that objectivity itself was a fluid concept framed by culture.[6] From the historical perspective this is not surprising: as a human endeavour science was always going to be framed by contemporary thinking and by personalities. Indeed, this idea is thoroughly embedded in current scientific method through the long-standing notion of 'eureka' moments.

This idea takes its name from the way Archimedes supposedly reacted when he hit upon the concept of water displacement as a means of identifying counterfeit currency.[7] The history of science is filled with such moments — as when 23-year-old Isaac Newton conceived of what he called 'fluxions', part of the mathematical language of calculus.[8] Most are mythologised: Archimedes probably didn't have the idea in his bath and then run naked through the streets of Athens shouting 'Eureka!' ('I have found it!').[9] Newton probably didn't discover gravity when an apple fell on his head, though his observations of apples falling from his tree in Woolsthorpe Manor likely had a part in the twenty-year process of developing his theory. Still, the underlying idea is true: flashes of insight arrive, unheralded and apparently without conscious thought. They are a well-known outcome of human thought processes. Rutherford was not shy in recognising it, as in 1932 when he told the Royal Academy of the Arts there was a 'strong claim' that 'the process of scientific discovery may be regarded as a form of art.'[10]

This is not to deny the worth of empirical data, or the effort to find objective ways of analysing it, or the quest for objective outcomes. On the contrary, data remains vital, and Rutherford himself was the doyen of objectively numbering everything. His world revolved around arrays of carefully recorded values, detected with his equipment and written down in notebooks and on paper sheets jammed into their bindings. His investigations into the structure of atoms involved literally counting individual subatomic particles.[11] It was not strictly empirical: it was possible to miss some of the dim and transient flashes that betrayed a reflected particle — a point seized upon by his critics.[12] The more crucial human part came in the *explanations* of what he saw. The first such experiments produced results Rutherford did not expect. To go from counting dim flashes of light to a conclusion that atoms were mostly empty space — thus radically different from the prevailing idea of the day — was a conceptual leap that Rutherford apparently made about half an hour before suppertime, one Sunday in 1911.[13]

This ability to conjure new concepts from fragments, all without conscious thought, appears to have been present throughout the human story. The ability of early humans to make stone tools is well known, indicating an ability to work in the abstract: they literally saw shapes in the rock. Evidence has been found of wood being cut for structural use by hominids nearly half a million years ago.[14] A later human type, the Neanderthals, developed epoxy resins possibly 190,000 years ago,[15] along with compound materials such as a mixture of bitumen and ochre which they fashioned into a non-slip grip for hunting spears.[16] Neanderthals are also believed to have made musical instruments and placed artworks on cave walls,[17] indicating abstract thinking. A cousin species, unnamed at the time of writing but informally called the Denisovans, had similar technologies and, implicitly, similar cognitive abilities.[18] Our own species, Homo sapiens, also showed such abilities at the time.[19]

In short, the thought processes blending abstract thinking, visualisation and creativity are integral with the human condition.[20] And one implication is that — irrespective of Western conceits of science as objective — the road to explanations is human and subjective.

This also means that such explanations are going to be wrapped in the idea-sets of the society in which the scientist lives. We can see that principle at work in one of the earliest fields of science. The Sumerian civilisation in the Euphrates river valley around 3100 BCE pioneered positional astronomy, seeing patterns in the stars and identifying the movements of the planets against them. This idea grew during the later Babylonian civilisation, whose priests believed that these movements predicted turns of fate in their everyday lives — a conceptual leap that survives today as astrology. The process was classic if-then logic: *if* Nergal [Mars] moves into Mul-Sipazianna [Orion], *then* event X will occur. While it is easy to scoff at such assertions today, it underscored basic relationships between observation, conclusion and the way that understanding is affected by wider social forces — in this case, the belief in higher powers and their influence on human activity.

The idea that there might be explanations not requiring higher powers was explored by the ancient Greeks, particularly during the classical period from about the fifth century BCE until the third. Four 'elements' were identified by Empedocles of Acragas— earth, air, fire and water. That was challenged by Leucippus and later Democritus, who proposed that all matter was instead made up of indivisible particles — 'atoms' — which existed in a void. Much of this was an outgrowth of philosophy. As Rutherford put it in one of his public lectures, the classical Greek view of atoms was 'pure theory, for it was not based on any definite scientific fact.'[21]

At the same time the Greeks pursued a hands-on approach to the world they could see and measure, making discoveries in mathematics and geometry that remained central to Western thought for millennia. The philosopher Aristotle, around 350 BCE, was able to prove that Earth was spherical by observing constellations as the planet rotated. A little later both Aristarchus and Eratosthenes produced measurements of Earth's radius. All were capped by Euclid (325–265 BCE), whose geometric principles and associated mathematics defined the spatial relationships of the everyday world. This idea-set was not displaced until the early twentieth century, when Einstein offered a new geometry for space and time, overthrowing Euclid's thinking.

Classical Greek civilisation influenced the subsequent Roman world, and many of the principles survived into the medieval society that then emerged across Europe. Trigonometry and geometry in particular made it possible for Europe's medieval engineers to build astonishing works of stone, such as cathedrals that ran to the very limits of what was possible given the tensile strength of their building materials.

Western science nonetheless wallowed in the millennium or so after the politico-economic collapse of Rome's Western empire in the fifth century. China and the Muslim world took the lead. Many of the ideas that emerged — notably the concept of the number zero — were later absorbed into Western thought. Others were re-invented. At the time, though, the West just drifted along. In large part the issue was social. Christianity gained power during Europe's so-called Dark Ages, with teachings that included doctrinal explanations drawn from the science of the ancient world. Other ideas circulated, including the concept of alchemy — transmuting one element into another, ideally gold. Alchemists believed this had been possible in the ancient world through a mystic material known as the Philosopher's Stone. Of course none of this worked, and eventually the idea of transmutation was widely discredited: it was simply magic, the province of charlatans and fraudsters. Elements were fixed entities, and that was an end to it. By Rutherford's time the idea of immutability was an absolute foundation of chemistry.

Europe's scientific fortunes changed with the Renaissance. The thinking that went with it reflected a change in style of thought — the paradigm shift that Kuhn showed also lies behind Western scientific revolutions.[22] The shift became faster in the fifteenth century as the arrival of printing presses in the West made information easier to obtain and dropped its cost. Literacy rates were still low, but an information age nonetheless followed. Among other things it helped drive a revolution in science that began in sixteenth-century Europe and the British Isles. Initial ideas came from thinkers such as Francis Bacon (1561–1626). To Bacon, understanding came from empirical data, which was obtained through experiments and assessed by reasoning. Bacon's process demanded another novelty, critical thinking. As he put it in his *Essays* of 1625: 'Read not to contradict or confute, nor to believe and take for granted, nor to find talk and discourse, but to weigh and consider.'[23]

Bacon was not the first to think along these lines. The Persian scientist Abu Ali al-Hasan Ibn al-Haytham (965–1040) had taken a similar approach 600 years earlier. However, the new paradigm gained pace across Europe during the seventeenth century, supported and given pace by the invention of new instruments to measure scientific parameters. Telescopes and microscopes emerged early in the century. The middle decades brought barometers, thermometers and pendulum clocks. The advent of these instruments forced scholars to pay attention to systems of measurement.

The telescope fuelled a major astronomical breakthrough by the mathematician Johannes Kepler (1571–1630) that has sometimes been considered the true beginning of Europe's scientific revolution, because it displaced the old Ptolemaic idea that Earth sat at the centre of the universe. This really was revolutionary in every sense of the word, not least because the Ptolemaic system was self-consistent, worked accurately, and fitted all the observed data. Kepler showed, nonetheless, that it was dead wrong. This point is worth remembering when we come to the way quantum superposition was explained by Rutherford's student, Niels Bohr. The story of Kepler's astronomical revolution highlights how paradigm change works in the Western scientific system, a process that remained true when Rutherford and his colleagues introduced their own new paradigms 300-odd years later.

Until the mid-sixteenth century, European philosophers, churchmen and scientists followed the geocentric model developed by Aristotle, Hipparchus of Nicea (c. 190–120 BCE) and Claudius Ptolemy (c. 100–170 CE). This put Earth at the centre, around which the Sun and planets circled along paths known as deferents. It was a complex system largely because, as seen from Earth, the planets occasionally reversed direction. This was particularly obvious for Mars, Venus and Mercury, which entered this 'retrograde' motion every 26, 18 and 3 months respectively. This was explained by means of 'epicycles', small circles centred on the primary path each planet took around Earth. However, further observations made clear that the epicycles didn't fully explain the way the planets were seen to move. So the epicycles themselves required smaller epicycles. Even that didn't fully explain things. That led to another element, the 'equant', by which the point around which everything revolved was slightly displaced from Earth itself. This last idea was first proposed by Hipparchus and then refined by Ptolemy.

The whole system was complex, but it was consistent and it worked. It could even predict eclipses: modern analysis has shown the error rate was about three percent.[24] Although Aristachus of Samos (c. 310–230 BCE) had proposed a heliocentric model of the solar system during the third century BCE, Ptolemy's geocentric concept became dogma.

But then in 1515 the Polish priest and polymath Nicolaus Copernicus (1473–1543) introduced a new paradigm. If Earth circled the Sun, most of the complexities of the Ptolemaic model were not needed. Copernicus found that retrograde motion was simply an illusion caused by Earth overtaking — or being overtaken by — any of the other planets. This still didn't *precisely* describe planetary positions, so Copernicus still had to allow for epicycles. He did not publish until just before his death in 1543, for fear of persecution, and his idea did little to dislodge the existing model. But cracks were appearing.

The full revolution came in the early seventeenth century. In 1610 the Italian scientist and astronomer Galileo Galilei observed Jupiter by telescope and discovered its major moons. He also discovered that Venus had phases like the Moon, showing that it was circling around something other than Earth. Ptolemy's ideas were looking rickety. Enter Johannes Kepler, official mathematician and astronomer for the Holy Roman Emperor, Rudolf II, in 1601. In 1609 Kepler published the first of two major works describing the movements of the planets in which he put the Sun at the centre and fixed Copernicus' problems by introducing one simple idea. According to Kepler, all the planets orbited the Sun in elliptical paths. This explained everything, removing any need for epicycles and offering a straight-forward model of planetary movement.

The importance of this story is the process. Humans have a relentless ability to see patterns and to apply them to what is observed. Ptolemy's geocentric model was just such a pattern. It worked correctly, and in terms of what was known at the time that made it credible. Then new evidence emerged to show that the idea was flawed. A new paradigm was then proposed. The initial effort didn't work, but the whole new set of ideas fell into place on the back of further work. This process — by which an accepted understanding falls into crisis, is replaced by a new idea, and where that idea is then refined — was identified by Kuhn as the way that Western science develops.[25] This was the way that the underlying paradigms of modern physics emerged at the hands of Rutherford and his colleagues.

By the second half of the seventeenth century a science revolution was well under way across Europe and Britain. Much of the science that emerged during these years came out of England at the hands of one man, Isaac Newton. To call him a polymath is perhaps an understatement: among other things he set up the basis for what became known as 'classical' physics, essentially Rutherford's starting point. Newton's timing was unsurprising: after a generation of chaos during which England had lurched from monarchy to civil war, dictatorship, and back to monarchy, life was becoming more stable. One of the outcomes was the rise of scientific institutions, notably the Royal Society. This was established in November 1660 by Sir Christopher Wren, the Professor of Astronomy at Gresham College in London. Robert Hooke (1635–1703) became the first Curator of Experiments in the sciences two years later. Then in 1665 the society began publishing the world's first scientific journal, *Philosophical Transactions*.[26] It swiftly became the most prestigious scientific institution of the age.

Newton, like Rutherford, was into everything. In a few dramatic years he co-developed calculus, the mathematical language for describing dynamically changing systems. He developed optics, building on and repeating the optical work of the Islamic scholar Ibn al-Haytham. Newton's experiments shining light into prisms showed that the Sun's light was actually composed of distinct colours, and much later this same principle was applied by Rutherford and his colleagues when they shone beams of X-rays into crystals to investigate their structure. Newton also defined much of what we now know as classical mechanics, including the principles of motion. Perhaps his greatest achievement was his theory of gravity, which he developed with input from Hooke and Edmond Halley (1656–1742) — *that* Halley, as in the comet. This was a towering achievement, and by 1686 Newton had his work ready in a three-part manuscript, *Philosophiæ Naturalis Principia Mathematica,* which so impressed the Society that the president of the day, Samuel Pepys, arranged to have it published.[27]

Newton's description of gravity was breathtaking, though it took more than a century for perhaps the greatest triumph to emerge. In 1845–46 the French astronomer Urbain Le Verrier (1811–77) and the British mathematician John Couch Adams (1819–92) independently used Newton's equations to analyse variations in Uranus' orbit. From that they deduced the existence of a major planet beyond. They were able to calculate where in the skies this new celestial body had to be, and in September 1846 the German astronomer Johann Galle (1812–1910) found it in less than an hour by pointing his telescope at the indicated position.

By the end of the seventeenth century, Newton and colleagues such as Robert Hooke, Edmond Halley, Robert Boyle, Christian Huygens, Gottfried Leibniz and others had altered the vision of Earth's place in the solar system. Thanks mostly to Newton, science had explained gravity, discovered that white light could be split into colours, and provided a way of both understanding and quantifying everyday motion. Other researchers, around the same time, identified the principles of gas behaviour such as expansion and response to heat, volume and pressure.[28] Furthermore, these discoveries were not merely for the elite. Printing meant that new ideas could be widely disseminated. They also attracted public interest. In England, particularly, science was popularised through another new phenomenon of the period: the public coffee house.

All this helped change the way European society framed its thinking, producing changes of philosophy that affected everything from social structures to politics. One of the more influential thinkers of this era was François-Marie Arouet (1694–1778), better known by the pseudonym he adopted in 1718: Voltaire. He was a fan of Newton and produced a wealth of writings that helped define the thinking of the century with its focus on reason and rationality.[29] Many of these concepts remain central to Western society. The emergence of this idea-set, along with the Industrial Revolution and the far-reaching socioeconomic shifts this brought, is why historians generally see the eighteenth century as marking the Western transition from an 'early modern' society to a broadly 'modern' one.

The new approach to science was highlighted by the way the absolute size of the solar system was measured in 1769–70. This embodied another principle that Rutherford exploited: sometimes a single piece of data could unlock an avalanche of consequential discovery. In the case of the solar system this process was elegant and surprisingly simple. By the late seventeenth century, further development of Newton's equations enabled *relative* distances to be calculated, typically using the Earth–Sun distance as a unit of measure. But the *absolute* distance in miles was the real prize. Newton's mathematics showed that if the absolute distance to *any* other body in the solar system was known, it would become possible to calculate the distances to *every* planet, from that calculate their true diameters, and find the mass of all planets with moons. Once mass and diameter were known, densities and surface gravities could be calculated in turn. In short, an avalanche of information would immediately follow once a single absolute distance could be found.

The problem was getting that first data-point to start the chain. Edmond Halley hit on a way in 1677, but did not publish his findings till 1716. The method involved observing a transit of Venus across the face of the Sun from two widely separated points on Earth. The height of the passage across the Sun's disc would vary, depending on where the observer was located on the Earth's surface. If both locations were known, Halley concluded, it would be possible to calculate the Earth–Venus distance by triangulation. He even worked out a way of measuring the height of the transit on the solar disc by timing it.[30] Halley knew he would not live to see the next transits, which were due in 1761 and 1769, but he made sure the methodology was available. Observations in 1761 failed to get results, so special attention was paid to the 1769 transit. By this time the British were taking an interest in the Pacific, largely to avoid being left behind in a fresh burst of Imperial rivalry with France and Spain. The Royal Navy captain Samuel Wallis (1728–95) reached Tahiti in 1767. The island was then selected as a suitable destination for a Royal Society expedition, organised with support of the Royal Navy, to observe and measure the crucial event in 1769. The expedition was led by Captain James Cook and secured the essential data for the Royal Society. With that in hand they sailed for New Zealand, establishing the region as a British interest in the South Pacific. That was a political bonus for the British government, but the key scientific haul was that crucial transit data.

The fact that astronomers had been able to calculate the exact times of a Venus transit, then use observations of it to obtain an absolute distance — so unlocking a vast amount of other information — underscored another side of the new approach to science. Everything was seen as a mechanism and the focus was on identifying the components and determining the relationships between them.

This concept continued to shape the way Western science developed during the eighteenth century. Perhaps the most enduring example remains the work of the Swedish naturalist Carl Linnaeus (1707–1778), who invented a system for categorising plants and animals. He published it in 1735, and Linnaean classification remains in place today. Indeed, classification remains with us today as a general organising principle. The upside is an ability to reduce the world to easily digestible patterns, something humans are hardwired to look for. These can be useful discovery tools. On the downside, such systems are sometimes confused with objective reality. Categorisations struggle when something crops up that doesn't fit the boxes. Rutherford's own work is a prime example: he saw himself as a physicist, yet, to his surprise, the Nobel Prize he received in 1908 for his work on radioactivity was awarded in chemistry. A more recent example is the 2006 creation of the category 'dwarf planet', to which Pluto was assigned as part of an attempt to scientifically categorise and define planets. The outcry across the science community at the technical inadequacies was followed by a louder scream from the public at the demotion of a much-loved social icon.[31] The whole issue was purely a human construction: Pluto, of itself, never changed.

The eighteenth century also saw strides in chemistry. One of the questions facing seventeenth-century scientists was why ashes weighed less than the wood they were created from. Something was clearly being lost in the combustion process, and as far back as the 1660s the polymath Johann Becher had proposed that it was due to yet another invisible and undetectable substance. He explained rust the same way: metals released this substance to the air, leaving rust as a residue. The magic substance was named 'phlogiston' in 1703 by Becher's student Georg Stahl, who derived it from *phlox*, the classical Greek word for flame. However, this began to look shaky when it was discovered that rusty metals actually *gained* weight, which made no sense if they were losing something.

Then, in 1774, the British chemist Joseph Priestley discovered that air was composed of a number of gases, only one of which was required for combustion. He explained this in terms of phlogiston, but an alternative was proposed by the French chemist Antoine Lavoisier: Priestly's combustible gas was being *absorbed* by metals, causing them to rust as they gained weight or mass. Phlogiston, in short, was an illusion. It took a while for Lavoisier's ideas to be fully accepted, but they helped to clarify the nature of chemical elements — substances that were the building blocks of other materials. Oxygen and hydrogen, for example, combined to produce water. These elements were themselves considered immutable, something that remained a paradigm of chemistry — until Rutherford showed otherwise.

In Europe, science also helped fuel the Industrial Revolution, a remarkable explosion of enterprise that began gaining pace in Britain from the 1760s, and which by the turn of the nineteenth century had provoked significant social and economic shifts. Thinking changed with it, and one of the phenomena that gained ground was the idea of progress. This assumed that change — any change — was directional: an objective improvement from what had gone before, advancing relentlessly towards an ultimate endpoint. In terms of the sciences that endpoint was complete knowledge — all of it carefully labelled, categorised, ranked and classed. The idea was given credence by the fact that engineering was on a learning curve, creating new technologies that ranged from steam engines to iron bridges, which became some of the defining features of the Industrial Revolution.

One outcome was that by the nineteenth century the notion of 'progress', as a process of objective improvement with time, had become embedded in Western thought. It was even reduced to a slogan that persisted into the next century — 'you can't stop progress'. Progress also became a mechanism for validating colonialism and judging the world. Peoples encountered by Europe's empire-builders were categorised as backward because they lacked Eurocentric trappings. The problem for Europe's scientists was that their discoveries were hijacked by this thinking, notably when the scholar Herbert Spencer (1820–1903) wrapped Charles Darwin's evolutionary theory with concepts of progress to create 'social Darwinism'. That helped fuel the social prejudices of the time and also provided a justification for period economic theory. Karl Marx, particularly, co-opted progressivism into his ideas.

Industrialised science also changed society. Industry enabled gas to be produced from coal in enormous quantities, piped to homes in growing 'middle class' cities. Gas lamps extended family life into evenings, fuelling markets for 'penny dreadful' novels, newspapers, and for hobbies. In 1856, 19-year-old William Perkin accidentally discovered that waste products from coal gas production could make vivid purple dyes, a colour that until then had been expensive and difficult to create.[32] This was extended to other colours across the spectrum. Suddenly city-dwellers could afford brightly coloured clothing that had been the province of the rich. Carbolic acid, another byproduct of gas production, was found to be a disinfectant: added to soap it became a commodity demanded by city-dwellers — an ideal product in an era when physical cleanliness was equated with moral purity. The world in which Rutherford was born and brought up was a product of these broad social trends, in part shaped by the achievements of nineteenth century science.

One question confronting scientists of the early nineteenth century was the nature of light. In the seventeenth century Newton decided light was made up of particles. His colleagues disagreed. Robert Hooke and the Dutch scientist Christiaan Huygens instead proposed that light was a wave. But if it was a wave it required a carrier medium. From this emerged the idea of luminiferous aether, an intangible and invisible substance that suffused everything and carried light. Newton wasn't happy: answering a question by imagining something invisible and undetectable stood against the principles of science. However, particles didn't explain refraction or diffraction either. So light had to be a kind of wave, and he had to accept that aether therefore existed. Others of Newton's day, including Huygens, were certain. Robert Boyle (1627–91), the Anglo-Irish philosopher, even tried to extend the concept of aether into explanations for magnetism and gravity.

This was no mere theoretical argument. The question of whether light was a wave or a particle reappeared in the late nineteenth century and was fundamental to the way modern physics emerged. But the consensus of the late seventeenth century was otherwise clear, and the concept of an invisible aether became one of the key frameworks of classical physics. Further discoveries merely reinforced the idea, as in 1801 when the English physician Thomas Young discovered that a diffracted beam of light produced interference patterns. This could only be explained if light was a wave. Further, in 1819 the French engineer Augustin-Jean Fresnel (1788–1827) proved the point mathematically. More experiments showed that light was a transverse or horizontal wave, as opposed to sound waves in air, or water ripples in a pond, which are longitudinal or vertical. And if light was a wave, it surely required a medium to transmit it. Clearly the aether

— invisible, intangible and undetectable — had to be real. It all made sense at the time.

Into this mix of discoveries plunged a further phenomenon: electricity. This was not strictly new. The classical Greeks knew that amber — in classical Greek, ḗlektron — could produce sparks by friction with another material. This received some attention during Europe's seventeenth-century explosion of science and was dubbed electricity by the physician Sir Thomas Browne (1605–1682) in 1646, drawing from a term coined by another physician, William Gilbert (1544–1603), who was in turn riffing on the Greek word. A few years later the German scientist Otto von Guericke (1602–1686) invented a device for producing an electrostatic charge. Investigations gained pace into the eighteenth century, buoyed by the invention of the Leyden jar, a device that could store a charge generated electrostatically — and which was crucial to Rutherford's early experiments in the 1890s.

The American scientist Benjamin Franklin (1706–1790) discovered that lightning was electrical in 1752. Then in 1800 the Italian scientist Alessandro Volta (1745–1827) invented the first electric battery, spurring new interest. At first electricity was popularly assigned near-magical powers. One result was that when Mary Shelley and her European tour party were trapped by bad weather in their Italian lakeside villa in 1816 she spent the time devising a tale by which electricity gave life. Today we know that her titular mad scientist, Dr Frankenstein, wouldn't have shouted 'It's alive!', as he did in the 1931 Universal Studios movie adaptation, so much as 'It's cooked!'[33]

It did not take scientists long to find out what electricity actually did. Many of the insights came from the British experimenter Michael Faraday. Like Rutherford, Faraday endlessly tested ideas by experiment. He initially focused on the relationships between chemistry and electricity, then moved on to define many of the basic principles of what became known as electrodynamics. The first inklings emerged at Faraday's hands when he discovered the relationship between electricity and magnetism, and postulated a force that extended beyond the capacity of his apparatus to demonstrate or study further.

This was a major turning point in physics. Electrodynamics, the science of the relationship between matter, electricity, and the forces involved, eventually came to dominate the field, though it was late in the century before it became a central theme. This was just in time for Rutherford to pick it up and run with it. Perhaps the largest shift came in the mid-nineteenth century when the Austrian physicist Ludwig Boltzmann (1844–1906) introduced methods of statistical analysis, feeding into the methodology Rutherford used a few decades later.

This focus on electrodynamics came amidst a structural shift in the way science was explored. By the late eighteenth century Western science was reaching the point where no individual could master all of it. A few still did, notably the English scientist John Dalton (1766–1844), whose work spanned everything from atomic theory to colour blindness and meteorology. He proposed that atoms were solid and indivisible, and in 1805 devised a system of atomic weights, defining hydrogen as 1 because it was considered the smallest atom, and everything else was measured relative to that value. This was not amended until Rutherford promoted the more useful concept of atomic number over a century later (chapter 9). Dalton, though, was an exception: for the most part, nineteenth-century scientists became increasingly specialised, and the sciences themselves became more clearly divided into categories. Physics, at the time broadly an amalgam of mechanics, thermodynamics, optics and astronomy, emerged as a specialty. Chemistry became a separate discipline.

The division between physics and chemistry was thought to extend to the smallest scale: atoms and molecules. Dalton's atom was regarded as physically solid and the smallest possible division of matter. The basic elements were composed of atoms, each with distinct chemical properties. By this time, however, it was accepted that atoms could combine in groups, called molecules. One outcome was that by the mid-nineteenth century molecules were considered to be the province of physics while atoms remained that of chemists. The point was underscored by the ongoing discovery of elements and efforts by chemists to explore just how and why chemical reactions occurred,

EXPLAINING IONS

IONS — ONE OF RUTHERFORD'S go-to tools during the early part of his career — were discovered by Michael Faraday in the early 1830s as a result of his experiments placing electrified plates or rods into various chemical solutions to create an electrical circuit that passed through the solution. This liberated material from the liquid, which then accumulated on the plates. At the suggestion of his colleague, the English scientist William Whewell (1794–1866), Faraday coined the term 'ion' to describe the material being moved electrically.

This experiment became standard in twentieth-century high school chemistry, usually involving copper sulphate. This produces a blue solution when dissolved in water. If two conductive rods are dropped into this mixture and connected to a power supply, current flows through the rods and into the copper sulphate solution, and pure copper is deposited on the negatively-charged rod. Meanwhile bubbles of oxygen appear on the positive terminal. This is known as electrolysis, the process of inducing a chemical change by electrical means. Faraday explored numerous chemicals, inventing the terms he needed along the way, including cathode (negative) and anode (positive). He dubbed the materials attracted to the positive rod anions, and those attracted to the negative, cations. Through this work he was able to demonstrate direct relationships between the mass of material liberated by electricity and the charge.

The problem for nineteenth-century scientists was that nobody knew what ions actually were. It took some time to piece the puzzle together. By the mid-1890s J.J. Thomson was exploring the nature of ions in gases, work that led to the creation of particle physics as a field of endeavour, and which also gave Rutherford a laboratory tool he used for much of his career. Thomson's work did not fully explain the mechanism, but everything fell into place just before the First World War when Rutherford established the nuclear model of the atom, in which atoms are mostly empty space but contain a physically small nucleus that holds the majority of the charge and mass, around which the electrons circle. It turns out that ions are atoms that have gained or lost electrons, meaning they have a negative or positive charge. It is also possible for molecules (multiple atoms) to have negative or positive charges.

Electrolysis was industrialised well in advance of this theoretical explanation. Patents to produce aluminium by electrolysis were lodged as early as 1886. This became the primary means by which the world's supply of some metals was subsequently produced, demanding enormous quantities of electricity to overcome powerful chemical bonds and reduce or break metal ores down to the elemental metal. The point is highlighted by the fact that a hydroelectric station with an eventual annual output of up to 4800 gigawatt-hours was built in New Zealand during the late 1960s, primarily to power an aluminium smelter that produced up to 335,000 tonnes of aluminium per year.[34]

The way that metal ions flow to the negative side of a circuit was also harnessed by industry as a way to evenly coat metals with a thin layer of another metal, a process dubbed electroplating. This is routinely used today for everyday products. The discovery that even plastics could be coated, via an intermediate conducting layer of copper, made it possible to give cheap consumer products a more solid look.

while by contrast physicists increasingly dealt with electricity and explanations of light and the aether. The overlaps between the two fields were less obvious. It was not until Rutherford's day that they began converging again, though even then, as he discovered himself when he got the Nobel Prize in chemistry for work he had considered was in physics, they were still seen as distinct entities.

The last decades of the nineteenth century brought another change in the nature of Western scientific research, buoyed by the growing scale of civilisation, wealth, and specialisation. The age of the 'gentleman scientist' gave way to a more organised approach built around formal qualifications and a growing scale of universities and institutions of learning. These sprang up across the United States, Europe, Britain and some of its larger colonies. And they did so just as populations grew and the number of students able to carry on with scientific careers expanded. Discoveries now became, more often than not, a product of joint work by a leading academic and their colleagues or students, often building on work done by similar teams elsewhere. This came about just as the new electrical technologies of the day provided instrumentation able to reveal new and previously unsuspected worlds.

All still pivoted on the inductive logic that emerged during the seventeenth century — a process by which hard data was collected, patterns sought, a hypothesis raised, and a theory developed. Testing the theory by further experimentation was crucial. Rutherford's own work followed this pattern. The curious point is that this predated the falsification method proposed by Karl Popper in 1959. By Popper's thinking, observation was by nature subjective. Inductive thinking did not lead to certainty. To Popper, the better way was to test every hypothesis against criteria that could prove it wrong. If the hypothesis survived such a test it could be considered robust. In some ways Popper's idea was a special case of inductive thinking, but it became a mainstay of later twentieth-century science. This had yet to be developed when the main foundations of Western physics were being laid by Maxwell, Thomson, Rutherford, Planck, Einstein,

Bohr and others. Yet even without this methodology, they turned the world of classical physics upside-down. Rutherford's explanation of radioactivity as natural transmutation, which we explore in chapter 6, is a case in point.

This, then, was the world of Western science as it had evolved by the time Ernest Rutherford was born: a Western-dominated field of human endeavour framed by the thinking of early modern Western society, built around mechanism, measurement, classification, and the conceit of objectivity. The shift of focus by the physics community to electrodynamics came as the field was professionalised by the industrialisation of science. Suddenly science gained pace. By the mid-nineteenth century the work being done in Britain's laboratories, in particular, was posing more questions than it answered. Something was in the wind, and that wind was rising to gale force even as Ernest Rutherford grew up in New Zealand.

3 | Young Rutherford

ERNEST RUTHERFORD WAS BORN AT SPRING GROVE, about 20 kilometres out of Nelson, in the northern part of New Zealand's South Island, on 30 August 1871. The local registrar of births, marriages and deaths recorded him as 'Earnest'.[1] Despite being described through his life and after as 'British', he always thought of himself as a New Zealander.[2] It was important to him: he never lost the grounding of his youth and the ordinariness of his upbringing. At a time when physicists were household names, elevated and uplifted as superior intellects, Rutherford took delight in presenting himself as a simple man close to the soil. 'Who was that Australian farmer?' one person asked a friend after meeting Rutherford for the first time at a formal dinner.[3]

The vision of Rutherford's colonial upbringing as one of simple surroundings creating a simple man, as his former colleague Edward Andrade (1887–1971) put it,[4] has gained further credence from the way Rutherford's life was mythologised after his death. His youth in a South Pacific colony was recorded in large part through memoirs written after he was famous. Some actively projected his greatness into his youth,[5] and in a sense this was not surprising. Even in the 1930s, when much of this material was written, great men were often portrayed as having been predestined for glory. This idea certainly leaked into the draft of one attempted Rutherford biography.[6] In such visions, great men began as great children, whose early lives were success stories that began a series of automatic steps upwards. Such thinking conjures up visions, perhaps, of a young Rutherford happily digging up radioactive ores or splitting atoms with a chisel: but they tell us more about period beliefs than they do about Rutherford.

Period biographies often placed Rutherford in an equally mythologised setting, a colonial New Zealand portrayed not as it was, but as idealised by mid-twentieth-century popular memory. According to this idea, Rutherford was born in the 'heroic and hardy pioneering days of New Zealand', his subsequent greatness typifying the 'noblest traditions of the young nation'.[7] Against these odds he had supposedly hauled himself up, according to one description, by 'sheer force of will and innate ability'.[8] Such visions set the direction for subsequent accounts.[9] Even Rutherford's student, colleague and official biographer, Arthur Eve, was not wholly immune to the imagery.[10]

Historians have since shown that nineteenth century New Zealand bore only passing resemblance to the way it was portrayed,[11] but the idea of a primitive frontier has persisted in some science-oriented literature about Rutherford. This is likely in part because the authors of such material take little time to dig into the evolving historiography of a small South Pacific nation at the edge of their story.[12] Even relatively recent biographies present Rutherford's youth as close to the soil, contrasting that with the sophistication of the physics he dominated.[13]

The practical realities of Rutherford's colonial environment and the relationship between that and his science were eventually tackled at academic level in 1981, when Lawrence Badash produced a paper for an Australian conference looking at the relationship between colonialism and attitudes to science.[14] Of necessity this was brief and relied on general New Zealand histories pre-dating the late-twentieth century renaissance of that field. But this study showed, for the first time, how the constrained scale of New Zealand's economy and society hampered the ability of colonial educators and scientists to engage with the higher end of the field.

This was the key issue. Historical work has since shown that the New Zealand in which Rutherford was brought up was more sophisticated, developed and urbane than legend usually admits.[15] The rugged pioneering environment often attributed to the New Zealand of Rutherford's youth was certainly true of his grandparents' world, and to an extent true too of some districts in the central North Island in Rutherford's day. But by the 1870s and 1880s colonial New Zealand's main urban centres and surrounding rural districts were well established, with modern facilities such as telephone and electricity appearing as fast as they could be paid for.

The reasons the Rutherford family settled in Nelson can be traced back to the earliest colonial days. New Zealand was drawn into Britain's ambit from the end of the 1760s.[16] However, there was little interest in the place, and the Colonial Office certainly had no intention of colonising. Officials were only driven to act at the end of the 1830s when the New Zealand Company, founded by convicted criminal, jailbird and dilettante social thinker Edward Gibbon Wakefield, launched plans to establish a colony.[17] Wakefield envisaged a privately run capitalist utopia built around instant cities, where a new and perfect society would emerge under corporate control. However, such a society outside reach of the law was anathema to the British, who hastily set up a Crown colony in New Zealand in 1840. Wakefield's adventure, meanwhile, failed dismally. The venture was under-capitalised, the wealthy industrialists Wakefield required never eventuated, and people didn't behave the way Wakefield's theories required. His solution was to persuade investors to part with yet more money and repeat the exercise. This led to another under-capitalised settlement in the northern South Island, which his company named after the hero of Trafalgar, Admiral Horatio Nelson.[18]

It was thanks to this that Ernest Rutherford's grandparents, George and Barbara, reached New Zealand in 1843. Nelson was a hodgepodge of prefabricated wooden buildings, unsealed roads and detritus with a look and feel typical of many places around the Pacific rim. The Rutherfords were among the last subsidised settlers brought out by the New Zealand Company, which was in financial trouble. George Rutherford had been brought out to assist with a sawmill at nearby Motueka,[19] but eventually settled in Spring Grove, about 20 kilometres southwest of Nelson.

The society that emerged from the ruins of Wakefield's flawed dream was more dimensional and evolved organically from the middle class ideals of Britain's Victorian age. It was socially urban: the word describes a mindset rather than a location. Even those living in rural areas had the social outlook of British middle class city-dwellers. This was matched by a soaring vision in which New Zealand was going to become a bigger and better Britain in the South Pacific. For a few heady years in the early 1860s, as the local pastoral economy flourished on the back of the gold rushes and demand for New Zealand wool, this even seemed possible. New Zealand was part of an expanding 'Pacific rim culture' built around the spread of Europe's settlers into Australia, New Zealand and the United States, buoyed by cross-Pacific trade, and given impetus by the lure of gold that drew tens of thousands of hopeful prospectors from California to Victoria and then Otago. But the money that flowed from a New Zealand wool boom that came with the US Civil War, then the gold boom that followed, did not last. By the end of the 1860s New Zealand's economy was floundering. In an effort to revive its fortunes the government borrowed money to build infrastructure, bring settlers in, and to give private enterprise more impetus.

These moves were gaining momentum when Ernest Rutherford was born, bringing sharp change with them: change that was given further power by the fact that — by nature — the colony itself was becoming socially more complex as a new generation emerged, giving a demographic depth that the colony originally lacked. And so the New Zealand of the 1870s and 1880s evolved from its colonial origins. Shifting alliances of wealthy pastoralists gave way to a world of elected boards and bureaucracies. By 1881, when Rutherford was ten, nearly 40 percent of the population lived in urban areas, making New Zealand one of the more urbanised societies in the world at the time. New social ideals emerged. Social militarism — the exaltation of military symbolism and its place in everyday society — gained pace, as did temperance. Women wanted the vote. New technologies were picked up with alacrity. Britain began electrifying in the 1880s, but so did New Zealand. Telephone systems spread, hot on the heels of the world's first such networks in the United States. Urban rivalries, framed by period ideas of objective improvement, helped drive these technologies into New Zealand at pace.

These dramatic shifts were underpinned by a surge of immigration, mostly from Britain, driven in part by government policy, in part by renewed problems in the mother country. One result was that by 1885 a significant percentage of colonists in New Zealand had arrived only in the prior ten years — over 144,000 of them, in fact, out of a total settler population of just under 570,000.[20] And a new generation was coming to power, settlers who had been born in New Zealand or who, like James Rutherford, had arrived while still children. The influx of British settlers from the 1870s helped fuel a new self-vision, given force by the rise of social militarism. New Zealand had failed to become a better Britain, but now it was the best of Britain's children. Settlers looked to Britain for validation — and with that came the idea that New Zealand would never be good enough. This had its sequel in the idea that anything achieved by individual New Zealanders could only be worthy if done in Britain. That idea is one of the key reasons why Rutherford became so iconic in the land of his birth.

New Zealand's national insecurity complex also masked the fact that the colony was well up with Western trends and technologies. At times it was ahead of the curve. The first trans-Tasman telegraph cable was laid in 1876, bringing New Zealand into near-instant communication with the rest of the world. Government made education compulsory in 1877, three years ahead of Britain. Universal suffrage was introduced in 1893, giving women the vote, again ahead of the mother country. Initiatives by a new Liberal government during the 1890s led to the colony being described as 'the social laboratory of the twentieth century'.[21] The real point of contrast between New Zealand and the mother country was in economic scale and structure. The colony drew most of its income from pastoral endeavours ranging from large-scale sheep farming to small businesses built around dairy farms, flax mills or sawmills. Many of these smaller enterprises were family-run, often diversifying with side businesses such as cropping, and it was here that the Rutherford family found its place.

By the early 1870s the Rutherfords were entering their second and third generations at Spring Grove. James, Ernest's father, married 23-year-old Martha Thompson in 1866. She was a schoolteacher from Taranaki. They lived in a small house built on the family property and had a large family of their own: Ernest was their second son and fourth of their dozen children. James Rutherford was deeply religious and a great believer in the sanctity of the Sabbath, not hesitating to beat his children with a birch rod if they so much as whistled on this sacred day.[22] Martha was a very different type, presenting as domineering but described as 'very just' in her expectations of 'full value' from her children, and woe betide any person who attempted sharp practices.[23] It was through this upbringing, perhaps, that Rutherford learned to snap at those who did not match up in his laboratory. His mother had also been a teacher, and as Ernest's brother Jim recalled later, 'on Sundays we would sit around the Broadwood piano purchased by Mother with her first earnings as a school teacher.' The children sang hymns — such as 'Abide with me' — along with a lot of Sankey and Moody melodies.[24]

Between them James and his brother John Rutherford ran several businesses, including a wheelwrighting operation. In 1872 they added a water-powered flax mill at nearby Waimea, stripping flax for its fibre, a raw material for rope. New Zealand was booming in the early 1870s, fuelled by a flood of new settlers and by government spending on infrastructure. But capital remained short, forcing them to find creative answers to problems that otherwise required cash. Everybody had to be multi-skilled. Later, Ernest Rutherford attributed his abilities as an experimentalist to the hands-on problem solving he had done as a child and young adult in his father's small industrial and farming operations. He certainly gained plenty of experience there. James Rutherford drew heavily on family labour, though not simply to save money. He always paid the boys for their work picking hops, among other tasks[25] — and even while still a child Ernest was part of the team.

James Rutherford moved his family to Foxhill in 1877, buying a 13.75 hectare farm adjacent to one owned by his brother-in-law. The property included a good nine-roomed house, a stable, two dairies and other out-buildings.[26] Later, Ernest's brother Jim recalled the house as being like a hotel.[27] About a third of the property was in bush. It appears the farm provided food for the family and a little income. James worked variously as a wheelwright, or contracting to the railways department where he helped build some of the bridges for the Nelson line. Ernest's jobs included, as his brother later recalled, helping supplement the household wood supply and milking cows.[28] Saturdays were 'real playdays,' Jim recalled, when he and Ernest would equip themselves with a huge meat pie made by their mother and venture forth for the day, often spending time at the Wai-iti River that formed one boundary of the property:

> *Some days we would go birds nesting and gathered a wonderful collection of eggs which my brother George ... would exchange for books and cash ... Other days we would be spearing eels and catching brook trout* [sic] *in the ponds in the bush. Then we had bows and arrows, slings and all sorts of shanghais and weapons for bird and beast ...*[29]

Ernest Rutherford's early farming life lasted only a few years. His father and uncle were on the lookout for new ways to expand their business, in James Rutherford's case largely because his growing family was expensive to clothe and feed.[30] Farming was part of a larger whole that included interests in a flour mill, a sawmill, a wheelwright shop and flax milling. Then in 1882 James and John Rutherford secured a ten-year lease on land at Ruapaka, near Havelock at the southwestern end of Pelorus Sound. Here they set up an industrial-scale flax-milling business, ultimately employing around 20 people to work on a steam-driven plant that processed around six tonnes of green flax daily.[31] James initially commuted between Havelock and Foxhill, but finally decided to bring his family across, moving there with Martha and their children in May 1883.[32]

Havelock was a former goldrush town. By the 1880s it was a small rural service centre. For the Rutherfords the appeal was the flax that grew around the area. Everyday life continued much as in Foxhill. As Jim later recalled: 'C.H. Mills and his wife were very good to us also the Brodies, the Readers … Scotto, Matthews, the Smiths [and] Doreens'.[33] Sundays, indeed, were intensely religious for the family: in the morning they attended church, in the afternoon Sunday school and church again in the evening.[34] The family were not sectarian and Jim Rutherford later recalled the pattern: possibly Anglican in the morning and Presbyterian at night, then again Wesleyan. He mused that his parents evidently thought that more ways than one led to Rome.[35] This upbringing influenced Ernest as an adult: he had a finely honed sense of what was called propriety, including objecting to women's décolletage.[36]

The move to Havelock came amidst the Long Depression, a lengthy economic downturn that has traditionally been seen by New Zealand historians as pure gloom. More recent economic analysis has suggested these doldrums were symptomatic of economic rebalancing, as settler society gained proper demographic depth and the economy gained both dimension and resilience. One feature was a flood of small and often family-run industries. The Nelson district — where the number of such enterprises rose from 134 in 1877 to 202 by 1885 — was also the most industrialised on a per-capita basis.[37] However, the raw statistics do not give an indication of the turbulence as businesses flourished and failed, something that certainly loomed large in the Rutherfords' experience.

An idea of the financial scale of the Rutherford family businesses can be gained from the fact that John Rutherford's Brightwater flax plant was burned out by a fire that cost him an estimated £2,000 in losses over and above insurance.[38] This equates to around $435,000 in mid-2020s New Zealand dollar values.[39] The scale makes clear that John and James Rutherford between them had access to capital, ability to employ workers, and were operating small industrial businesses. George Rutherford — the eldest of James's sons — left Nelson College in 1883 and was fully employed in the family business. His diary reveals nine-hour days and wages of £6 14 shillings a week.[40] This was a significant sum at a time when unskilled wages averaged just under £2 per week.[41] George spent ten shillings of it on a watch — about $118 in mid-2020s New Zealand terms.[42]

In short, the Rutherford family were not among New Zealand's poor. However, capital value did not necessarily translate to turnover, turnover was no guarantee of profit, and flax prices were volatile. James Rutherford's problem was turning a steady income. At one point in 1884 prices were so low that James and John put the operation up for lease.[43] They were able to restart production in September, but in mid-1885 the market collapsed again. This time they had no reserves. The brothers turned to sawmilling, primarily to make railway sleepers. George was employed to run the operation, using the steam plant previously used for the flax mill.[44] James still had his land at Foxhill, with its hop field.[45] Available documentation, much of it collected after Rutherford's death, does little to reveal the emotional impact on Rutherford at the time. But his subsequent quest for financial security as an adult — including a reluctance to change jobs — suggest the experience left its scars.

The 1880s were difficult years for the Rutherford family in many ways. Ernest's youngest brother Percy died of whooping cough aged just 12 months in 1883.[46] Then tragedy struck in January 1886 when Herbert, Jim and Charles were taken boating with a group of others on Pelorus Sound, only for the small boat to capsize. Herbert and Charles were drowned.[47] Ernest would have been with them had he not been sent to the flax mill on an errand. Martha was playing her Broadwood piano when the news reached their Havelock home. According to family sources she never touched the instrument again and became a sterner figure.[48]

It would have been unsurprising if Rutherford's education had suffered as a result of these tragic events and the wildly fluctuating financial state of the family. But he was lucky: James and Martha were determined that their children should be properly educated. Ernest was initially taught at home by his mother, herself a schoolteacher, then sent to the nearby Foxhill school.[49] He did not do well at first: his school nickname was Dopey, and he dribbled.[50] But he showed a flair for mathematics. At home his hobbies turned to construction: he built a camera and used it to take photographs of his family. He took clocks to pieces to find out how they worked.[51]

At age ten Ernest was given a textbook written by Balfour Stewart, professor of physics at Victoria University of Manchester.[52] Stewart's 149-page text covered everything from gravity to states of matter, properties of solids, liquids and gases, the physics of moving bodies, sound as vibration, the effects of heat, and information on electricity as was known then.[53] Stewart urged enquiry, which he called a 'mode of questioning nature ... called experiment.' He urged his readers to try things for themselves, for example demonstrating laws of motion with the help of a pot of peas which — as he explained — were going to end up scattered about on the floor.[54] It was the kind of mischievous outcome likely to appeal to children. Such experiments were described and illustrated through his book, illuminating the basic principles of classical physics. Some of his examples involving pressure and mechanism were integral to the machinery operating in the various Rutherford family business ventures.

Stewart's text also explained how to calculate the velocity of sound in air by measuring the interval between seeing a cannon flash and hearing the report, over a precisely known distance.[55] As more than one of Rutherford's biographers have noted, Rutherford's father later found him calculating the distance of a thunderstorm.[56] Less often noted is the fact that Stewart's book included a detailed description of a device designed to measure the frequency of sound.[57] The concepts engaged with Ernest's interest in mechanisms.

The move to Havelock brought a new teacher, 27-year-old Jacob Reynolds. Havelock School was a typical small town institution of the day, founded in 1861.[58] By Rutherford's time it had around a hundred pupils,[59] and Rutherford gained a new nickname: 'Windy'.[60] After his first weeks the school began operating from a new purpose-built building formally opened by Prime Minister Frederick Whitaker. Nothing about this was special: there were many similar schools across New Zealand at that time. Rutherford was also not the only Havelock pupil to gain international recognition in science: another was William Pickering, later head of the Jet Propulsion Laboratory for 22 years from 1954, whose 1936 doctoral thesis drew from the principles Rutherford had established and applied them to cosmic radiation.[61]

Rutherford was reportedly quiet, with little interest in sports — by twenty-first century standards he was an archetypal nerd. Later he credited Reynolds with introducing him to the mysteries of Latin, algebra and Euclid.[62] He read widely, taking a keen interest in Dickens.[63] Where next was clear: both James and Martha wanted their children to be well educated, so that meant secondary school. However, this was not free in those days, and the family had little cash to spare. George had been to Nelson College on an Education Board scholarship worth £40 a year. It took Ernest two attempts to win his own scholarship, but he started as a boarder at Nelson College in early 1887. His performance in the second attempt at the scholarship exam impressed the headmaster, William Justice Ford, who started Rutherford in fifth form. As his fellow student Charles Broad said, this was an unusually high form for a new scholarship holder. 'We, the gods of the fifth form, wondered who this fair-haired interloper was. We soon found out.'[64]

Nelson College was a boys' boarding school, founded in 1856 and styled on the British model with its pseudo-military focus on personal discipline, tidiness and a regimented life. Even the main buildings evoked British schools.[65] However, this colonial version was not identical to the British model. New Zealand settler-age society was drawn from the middle classes and took pride in de facto egalitarianism, giving a unique local twist. By contrast with Eton, Harrow or other British public (meaning private) schools, Nelson College was a small institution — something that worked to Rutherford's advantage because it gave him better access to his teachers.[66]

One of the tropes of Western history is that outstanding minds always struggled at school. Thinkers of Rutherford's era who had this problem classically include Albert Einstein, Samuel Clemens (Mark Twain), Winston Churchill and Thomas Edison. Such people were easily able to develop new ideas and concepts, but often could not express them clearly in words. A remarkable number of scientists fall into this category. Some were socially awkward, some struggled to stop letters or numbers on a page dancing into an incomprehensible jumble, some wrestled with physical co-ordination to the point where tying shoelaces was a challenge, or — in the case of J.J. Thomson — gained repute for clumsiness to the point where they were not allowed to handle delicate laboratory equipment. Another common symptom was having the idea clear in the mind but not being able to put it clearly into words. This was disastrous at school: to teachers the pupil's use of a wrong word meant they didn't know the right answer, not that they'd suffered a cognitive tangle.

In general, such students were an ill-fit to a nineteenth-century education system that defined success in terms of rote-memorising letters and numbers and being able to mindlessly chant them back. Anybody who couldn't was dismissed as too stupid to learn, deserving only ridicule and beatings. If being beaten did not incentivise the child to just snap out of it and become normal — well, that was their own fault and they deserved all the pain and suffering the teacher could deliver. Small wonder that many remembered the experience as being 'menaced' with education.[67] Was Rutherford in this category? Certainly he struggled to express himself despite clearly having the right idea in mind. Wrong words occasionally came out, too. But he had few difficulties fitting into school.

At this stage Rutherford had no clear vision for his future, but he had been instilled with a work ethic and studied hard. In December 1887 he won the Stafford scholarship in English history, worth £20 annually for three years, and a Senior Classical scholarship worth £20.[68] These prizes were useful. Annual board alone was £50 and money remained tight for James Rutherford, who ran into fresh financial trouble when the government stopped buying railway sleepers from district suppliers. A local return to flax processing was not an option: supplies around Havelock were largely cut out. John Rutherford had already moved to the North Island, where he founded a new flax mill. James decided to follow, finding a substantial supply of flax on the Taranaki coast at Pungarehu, in an area only recently made available to settlers. In October 1887 he put the Foxhill property up for what he called 'low rent to a good tenant.'[69] Nobody was interested, and he was still advertising it in December.[70] In January 1888 James put the property on the market for outright sale. Even then there was little interest.[71] But that did not prevent James taking the family to Pungarehu in early February that year. They took the flax-processing plant with them and by April were back in business. Prosperity was rising and James was able to employ around 20 hands in the factory.[72]

Rutherford went on to sixth form that year, where by June he was first in classics, had worked up his English exercises in 'capital style', and was a 'very quick and a very promising mathematician … easily first', according to his maths teacher, the Scottish-born William Littlejohn.[73] Rutherford recalled Littlejohn as a strict disciplinarian but also a fine teacher of mathematics whose lessons included basic science. Rutherford was little interested in the science but, as he put it later, 'thoroughly enjoyed' the mathematics.[74] By December his progress in maths was so rapid that Littlejohn feared he might not have quite assimilated all of it. Ford, summing up Rutherford's mid-year report, said he was 'top in every form and his conduct is irreproachable.'[75] That year Rutherford won the Senior Literature scholarship, worth £12.10.[76] Littlejohn coached Rutherford during the years he spent at the college, adding him to after-school bonus classes.

Rutherford continued to study in Nelson College during 1888, taking the steamer north to Taranaki during the holidays where, on the first such break, he engaged in some fruitless shooting, then challenged his brother Jim to a boxing match and, as Jim later put it, 'cleaned me up in style.'[77]

Two years of high school made Rutherford one of the more educated New Zealanders of his day. He had the option of becoming a teacher himself, and apparently looked into it. But there was also the possibility of taking his education further. He could get to university by passing a matriculation exam. University study was rare at the time in New Zealand and relied on either being independently wealthy or winning a scholarship. The latter was available by passing a far more challenging exam than a matriculation test, but gave access to a Junior National scholarship, with three years of financial support.

Financing his son's tertiary education was a problem for James Rutherford. His prosperity was rising in Taranaki — but he was also having a new house built and pouring available cash into his business, building it into a multi-mill enterprise in partnership with his son George. Ernest took the scholarship path, swung, and missed. He was always nervous in exams, and it turned out that being top of his school did not translate into national status. He came twenty-sixth, but needed to be tenth or better to get the money.[78] And so he returned to Nelson College in 1889, where he won the Simmons prize in English literature, worth £6.[79] During 1889 he surged ahead in science: his texts included a university-level publication, *Sound and Light*.[80] He was made dux, a term derived from the Latin for 'leader'. That got him the nickname 'Quacks', which led to his youngest brother Arthur following him about their Pungarehu house making duck noises.[81] Rutherford sat the University Scholarship exam a second time in five subjects: Latin, English, French, mathematics and science.[82] He did not expect to get it and applied for work at

New Plymouth High School but — somewhat to his surprise — he was placed fourth nationally.

Rutherford selected Canterbury University College, registering there as Student No. 338.[83] He was enrolled for an arts degree, but he was brimming with enthusiasm for the new science — arriving at university just as physicists were realising that the discoveries of recent years were posing more questions than could be answered.

4 | Electrodynamics at Canterbury

BY THE TIME RUTHERFORD reached Canterbury University College, physics was evolving in extraordinary ways, buoyed by the hot topic of the era: electrodynamics. This emerged as a new field during the mid to late nineteenth century and by 1890 was well on the way to becoming the new 'operating system' around which physicists understood how the world worked. Rutherford encountered this head-on during his postgraduate studies at Canterbury, where he explored the effects of alternating current electricity on metal. This profoundly shaped the way he then approached physics, giving him techniques and styles of thinking that he applied for the rest of his life.

The connection between electricity and magnetism was first recorded as early as 1820 when the Danish scientist Hans Ørsted (1777–1851) noticed that a compass needle could be affected by an electric current. In other words, electricity had a magnetic component.[1] The following year the French scientist André-Marie Ampère (1775–1836) found that two parallel wires exerted a force on each other. The direction of this force was found to be at right angles to the direction of the current. Ampère dubbed this new phenomenon 'electrodynamics'. It became the in-topic of the day, keying into the mood of the era with its flood of new inventions and the concept of progress.

One early champion was Michael Faraday, who put a great deal of work into exploring the forces associated with electric current. Like Rutherford eighty-odd years later, Faraday was a hands-on experimenter, similarly driven to push the boundaries of knowledge with a fervent energy. He demonstrated that magnets produced a field with distinct lines of force — something anybody can repeat today by sprinkling iron filings on a piece of paper and bringing a magnet up against the other side. Then, in 1831, he was able to show that a moving magnet induced an electric current and, similarly, that a moving electric current produced a magnetic field. From these discoveries he was able to build the first electric motor and its inverse, the first direct current (DC) generator.

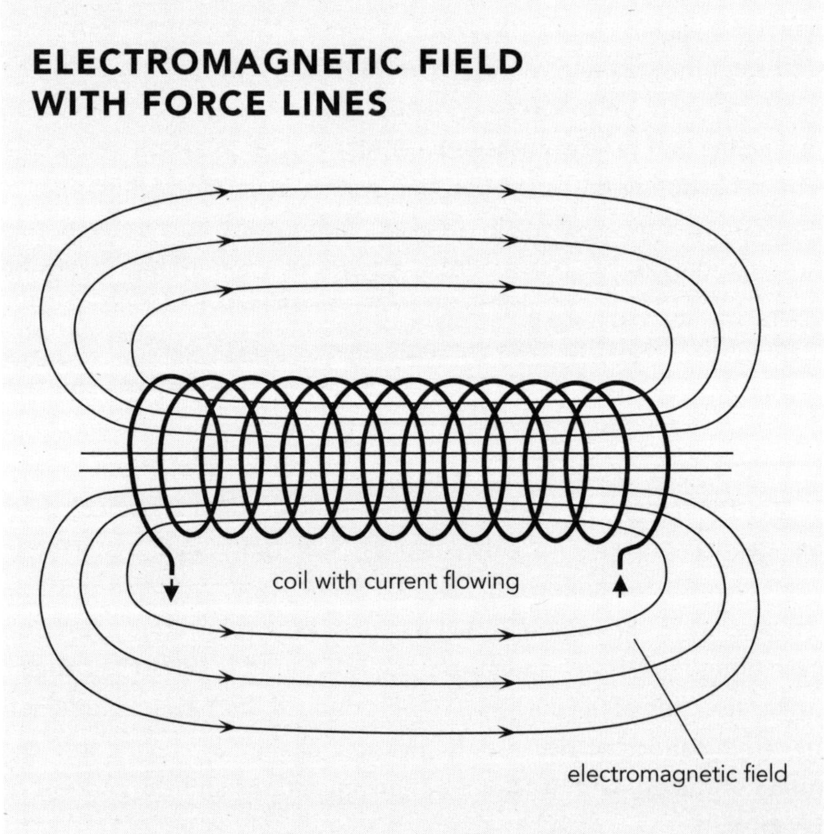

Unlike Rutherford, Faraday had little mathematical knowledge, and it was up to others to wrap his discoveries in this language. Meanwhile, scientists elsewhere in Britain and on the continent pursued investigations of their own into the same phenomena. In the 1840s James Joule (1818–89) began work on the way electrical current heated conducting materials, a theme also taken up by others including William Thomson (Lord Kelvin, 1824–1907), Gustav Kirchhoff and George Stokes. Then, in 1857, Kirchhoff built on the work of Wilhelm Weber (1804–1891) and Rudolf Kohlrausch to show that electricity propagated down a wire at nearly the speed of light.

Electricity, magnetism and light were clearly connected, but that needed proof. The Scottish physicist and mathematician James Clerk Maxwell (1831–79) picked up the sceptre in the 1850s. His first paper, published in 1856, looked into Faraday's force lines, declaring that empirical testing was the only way forward. He had already observed that the inverse square law applied to Faraday's magnetism, concluding that the principle was identical to the one already observed in the transmission of heat.[2] That same year, the German physicists Wilhelm-Weber and Rudolf Kohlrausch (1809–1858) identified the ratio between a static electrical charge and a dynamic one. Maxwell plugged that in to Newton's equation for the velocity of sound and came up with an answer very close to the speed of light, as then known. Maxwell concluded from this that electromagnetic radiation was a form of light.

This was only the beginning of a careful process that allowed Maxwell to define magnetic force lines in more detail, culminating in his seminal 1864 paper, 'A dynamical theory of the electromagnetic field.'[3] This was the key moment. In 53 pages of printed text featuring 108 numbered equations and a lot of unnumbered additions, Maxwell showed that electricity, magnetism and light were all aspects of the same force, which propagated as polarised waves. He was able to define this with 20 specific equations, which he referred to as the 'General Equations of the Electromagnetic Field.'[4]

With this one step, Maxwell unified recent discoveries into a single framework and gave them a mathematical basis. This was as revolutionary and ground-breaking for its day as Newton's theory of mechanics, this time applied to the dynamics of electromagnetism. His treatment remains one of the more remarkable single insights into the nature of physics, standing alongside Newton's *Principia*, Rutherford's work on radioactivity, and Einstein's work on space-time. In a single proposal, Maxwell unified existing thinking about apparently disparate phenomena and offered a way of understanding them.

Maxwell clearly had Newton in mind, because he also questioned whether gravity fell into the same category. Here he stalled on the fact that, unlike electricity and magnets, gravity had no polarity. It simply attracted. That stood in contrast to electromagnetism. Electrical charges could be negative or positive: opposite charges attracted and like charges repelled. Magnets had a north pole and a south pole; and once again, opposite poles attracted and like poles repelled. Gravity, it seemed, was something different. However, Maxwell's failure to produce a 'theory of everything' did not reduce the importance of his theory of electrodynamics. Even at the time his sweeping explanation was extraordinary. In the fashion of the day he presented his concepts initially as a talk, publishing the following year in the Royal Society's *Philosophical Transactions*.

Maxwell's 1864 paper was arguably *the* founding document of modern physics, kicking off and shaping the electrodynamic revolution that followed. But nobody noticed at first. As Einstein observed later, it took physicists some time to grasp Maxwell's conceptual leap,[5] and in other ways Maxwell's thinking was firmly of its era. He couched his ideas in terms of aether, concluding that electromagnetic forces were transmitted through this medium.[6] There was nothing unusual about this: new discoveries were always presented in terms of existing frameworks. And at the time there was little to suggest that this invisible, intangible aether was not real.

The developments that followed were also given shape by the other main development of the mid to late nineteenth century: the way scientific research transitioned from a 'gentleman hobby' in which talented individuals worked largely alone, into an organised and larger-scale enterprise based around research institutions and universities. This created the institutional world in which Rutherford worked. It was a product of both the scale of society and of its wealth. The phenomenon was shared across the Western world, notably the British Empire, Europe and the United States.

One of the first major institutions to emerge was the Cavendish Laboratory at Cambridge University. This had origins in 1869, driven in part by talk of Oxford University setting up a science faculty facility — something that Cambridge, naturally, had to match. The administering Senate organised a committee to look into the idea but were stalled by lack of funds. However, the chancellor of the day, William Cavendish, Seventh Duke of Devonshire, had a bent for science and provided £6300 to set up a laboratory to pursue experimental physics. Just who might run this facility was debated. Major names in the field such as John Strutt, the 3rd Baron Rayleigh (1842–1919) were considered, along with Lord Kelvin (William Thomson). In the end Maxwell was appointed the first professor of experimental physics in 1871. The facility opened in 1874 in a building on Free School Lane in Cambridge.[7] But Maxwell did not survive to run it for long: he died in 1879, and Rayleigh stepped in.

Rayleigh was another colossus of his day who, among other things, used Maxwell's electrodynamic equations to show why the sky is blue. The principle was dubbed 'Rayleigh scattering' in his honour. But this was perhaps the least of Rayleigh's ideas. His discoveries in acoustics included the principles behind stereo sound. He helped develop fluid dynamics — the physics of flow — an area spanning a vast range of human activities, defining everything from the design of ships and aircraft to pipe diameters required for efficient sewerage systems. Perhaps more crucially, Rayleigh worked with James Jeans to identify the principles behind black-body radiation, something that set things rolling towards quantum mechanics and the Rutherford-Bohr atomic model that we explore in chapter 8. Rayleigh held the post for five years before handing it to another rising physicist of the day, Joseph John

Thomson (1856–1940), 'J.J.' to his friends and colleagues. These three — Maxwell, Rayleigh and Thomson — together made the Cavendish one of the world's leading centres for experimental physics.

The British were not alone. French and German institutions expanded during the same period, part of the broad 'bureaucratisation' of the Western world, helping drive physics as a new field across the continent. And it was here that the first proofs emerged of Maxwell's theory, mainly at the hands of German physicist Heinrich Hertz. His background was framed by the mathematical work of Franz Neumann, Wilhelm Weber and Gustav Kirchhoff, all of whom looked into electromagnetics during the middle decades of the century.[8] In 1886 Hertz discovered that a coil set up with a small gap in the wire — producing a spark — could induce a similar spark in an adjacent coil, without either touching. In short, electromagnetic energy was travelling from one coil to another, through the air. He had discovered what we know today as a spark-gap transmitter. At the time its mystery remained to be unravelled, but it was a practical demonstration of electrodynamic forces at work, just as Maxwell's mathematics indicated.

Hertz spent the next two years exploring the phenomenon, demonstrating that these waves behaved in the same manner as light. But then things got strange. In 1887 he discovered that shining an ultraviolet light across two bare wires altered the voltage at which sparks began jumping between them. He called it the 'photoelectric effect'. This demonstrated that light was electromagnetic, just as Maxwell predicted. But there was a problem: the effect didn't behave in the ways expected if light was a wave. What was going on?

Into that mix flowed the question of the aether. Empirical proofs of its existence were becoming increasingly urgent given the superstructures being built atop this one assumption. There was some hope that the aether might be identified by indirect measurement. In the 1840s the British physicist George Stokes had argued that the aether behaved like a normal fluid when a body such as Earth moved through it, resulting in a measurable 'drag' effect.[9] This argument became widely accepted. Thanks to Kepler's laws of orbital

ELECTROMAGNETIC SPECTRUM

THE FIRST PART of the electromagnetic spectrum to be analysed was light, but this was just one octave out of the total comprising the whole spectrum, running from low-frequency radio waves through microwaves, infrared (heat), visible light, ultraviolet, X-rays and gamma radiation.

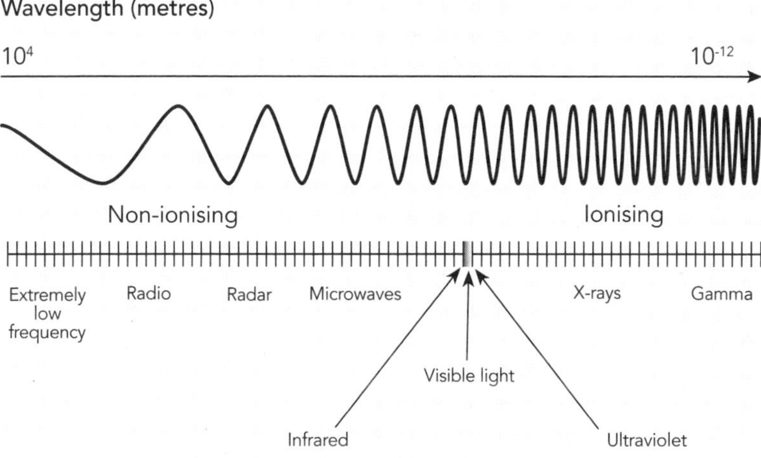

mechanics and Newton's theory of gravitation, Earth's velocity as it orbited the Sun could be calculated.[10] In practise this velocity varied slightly because the Earth's orbit is slightly elliptical, and the figure as known in the nineteenth century has since been refined. But overall it averaged about 29.79 kilometres per second. This was thought to be fast enough to create a 'headwind' in the aether, slowing the speed of light in the direction Earth was travelling. This was entirely reasonable according to the classical principles then thought to underpin electrodynamics.

The main challenge was making that measurement. The headwind from Earth's motion was minuscule when set against the speed of light. Maxwell thought it might be measured by timing the eclipses of Jupiter's moons, and wrote to the US Nautical Almanac Office in 1879 to ask whether that could be done. This drew the attention of a young naval officer working at the facility. Albert Michelson (1852–1931) was a German-American who redefined the velocity of light while a science instructor at the US Naval Academy in Annapolis.[11] Now Michelson wondered whether measuring the velocity of light in different directions might do the trick instead. If that velocity varied in the direction that 'aether drag' operated then it would show that drag existed, proving the aether was real.

To do this Michelson envisaged a specially designed interferometer, a device designed to split a beam of light into two, recombining them at an eyepiece. This had long been known to create an interference pattern, visible as bands of light and dark stripes. Michelson reasoned that an interferometer able to split one beam off at a 90° angle would enable him to test aether drag. If one of the beams was slowed it would create a 'fringing' effect on the interference pattern. The scale of this fringe could be predicted mathematically and the practical result measured against that prediction. This would prove by hands-on experiment that light was slowed in the direction of Earth's movement and thus that the aether existed.

Michelson completed the device in early 1881 while working in the Astrophysical Observatory in Potsdam.[12] But he failed to find a result. This could simply have been due to the machine not being sensitive enough. Rayleigh, by this time in charge of the Cavendish Laboratory, urged Michelson to try again. It took Michelson some years to develop a better instrument, this time with the help of the chemist Edward Morley (1838–1923). The new device was built around a stone block floating on mercury in a cast-iron trough, enabling it to be revolved smoothly between 16 markings built into the edge of the trough. It was ten times the size of the earlier interferometer, giving considerably greater fidelity. Michelson and Morley first used this device in April 1887, taking many measurements at differing times of day. They also considered the possibility that the motion of the

Sun — which added a velocity vector to Earth's own motion — would distort the answers. The velocity of the Sun had yet to be measured. However, by repeating the experiment after three months, when Earth had completed a quarter-orbit, Michelson and Morley were able to eliminate this factor.

This second experiment was a thorough attempt to discover aether drag by invoking all the variables they could build in and systematically investigating the question by empirical test. They found nothing. The speed of light, it seemed, was identical in all directions, irrespective of Earth's own motion.[13] That was a problem. Either the aether was at rest relative to Earth — which didn't make sense — or it didn't exist. But if it didn't exist, how could light and all the other electromagnetic wave phenomena be transmitted?

All these developments — the discovery of electrodynamics, the arrival of heavy duty mathematics, a new raft of questions and directions, the expansion of the institutional side into organised universities and research laboratories, and its effective professionalisation on the back of that expansion — laid the direct foundations for Rutherford's work.

What first captured Rutherford's enthusiasm were the questions raised by the way electrodynamic theory was being applied to new technologies: specifically, high-frequency alternating current and its slightly accidental child, radio transmission. Much of this was a direct extension of prior theoretical work by British and European physicists, notably Hertz. But it also had a good deal to do with gadgetry, giving it a clear appeal to Rutherford who retained his enthusiasm for clocks and machinery.

The lead was taken in the United States, where science underwent similar transformation of scale from the late 1860s, but where theoretical physics took second place to hands-on application for financial gain. The classic case was the way that the American inventor Thomas Edison (1847–1931) industrialised the process in 1876, setting up a laboratory to develop commercially saleable technologies. The results included a better telephone (1877), a sound recording and playback system called the phonograph (1877), incandescent filament light bulbs (1879), and motion pictures (1889). Other inventors in the United States, Britain and Europe found themselves having to skirt Edison's patents.

Edison also took a lead in providing power supply networks during the early 1880s. What became known as the 'battle of the currents' followed. Edison's direct current (DC) was easier to generate, but alternating current (AC) was easier to transform to different voltages and send over longer distances. The way around the AC generation problem was a device known as a commutator, a rotating switch that reversed the current of a DC generator every half-turn. Multiple inventors lodged patents for commutators during the early 1880s. A system for handling high loads without arc-over was then developed by former Edison employee Nikola Tesla (1856–1943).[14] He was Hungarian and had trained in mathematics and physics at the Technical University in Graz, then at the University of Prague. Some of Tesla's inventions remain in use today, including three-phase induction, used in everything from electric power tools to generators.[15]

Tesla decided he could exploit Hertz's spark-gap to send power through the air via magnetic fields oscillating at high frequencies. He imagined transmitting this to paying customers who could pick it up via receiving coils tuned to the broadcast frequency. There was only one problem. The energy transmitted fell off dramatically with distance. This was a fundamental principle of electrodynamics, and Tesla knew it. He tried using tuned intermediate boosters and receivers to offset the issue, but the laws of physics couldn't be bent.

Still, there was something else this phenomenon could do, and at far lower levels of power — levels so low they were hard to detect. The energy being broadcast by these methods could send information. All that had to be done was to switch the oscillating magnetic field on and off, producing pulses of energy. Messaging by stop-start pulse codes was how wired telegraph worked, sending that energy down the wire.[16] But that wire was also the limiting factor. Cables were expensive and limited the system to communications hubs: for example ships at sea had to call into port to pick up messages. By this time long-distance rapid communication had become part of the global social and economic fabric, vital to long-distance communications, to diplomacy, to controlling rail systems and more.

This was where Hertz's discovery came in. He never saw a use for spark-gap transmission, which he envisaged purely as a proof of Maxwell's theory, without practical application.[17] Others begged to differ. Energy was travelling between the spark-gap transmitter and the receiver. The inverse square law meant that signal strength rapidly dropped to negligible levels, but in 1890 the French physics professor, Édouard Branly (1844–1940), came up with a way of detecting these minuscule energies via a tube of iron filings. But much remained to be discovered to make such things work efficiently. Electrodynamics remained in its infancy and at the time even the efficiency with which various metals conducted electromagnetic forces was not precisely known.

This was a fertile field for any inquisitive young mathematician and physicist, and all this was unfolding when Rutherford left Taranaki in February 1890, heading for Christchurch and his first year at university.

Canterbury College was a small institution, evolving in 1873–74 from a Collegiate Union founded in Christchurch to further higher education. It became part of the University of New Zealand and by 1890 had seven professors and around 150 students.[18] By the early 1890s it was operating primarily from stone buildings opposite the Christchurch Museum, designed to echo British establishments such as Oxford. Science fell to Alexander Bickerton (1842–1929) — 'Bicky' to his students. He had emigrated to New Zealand with his family in 1874 to take up the post of Chair of Chemistry at Canterbury College and was viewed by some of his colleagues as a left-leaning eccentric. His prominent students of the era included Ettie Rout, an activist who became a controversial national figure during the First World War.[19] Bickerton became Rutherford's tutor, mentor and ultimately friend. Another student, Robert Laing

ELECTROMAGNETIC INDUCTION

THE PRINCIPLE OF electromagnetic induction was discovered by Michael Faraday, who realised that if a magnet is moved inside a wire coil it creates a current in the wire — and vice-versa. That is why alternating current (AC) electricity produces a magnetic field: the current is constantly reversing direction. The oscillating magnetic field, produced at right angles to the wire, can in turn induce current in a conductor some distance away. That principle sits at the heart of many twenty-first century technologies: everything from radio (including everything from broadcast radio to wi-fi) to electric motors, generators and more.

The problem with such a system is that transmitted energy falls off according to the inverse square law, in which doubling the distance results in received energy over any given area dropping to a quarter. For this reason a system able to broadcast useful energy the way Nikola Tesla hoped it might was not possible.

Household devices that exploit this principle include induction cooktops. These produce an oscillating magnetic field which *induces* heat in saucepans placed atop the emitter. This heat comes from two sources: electric eddy-currents that dissipate into heat, and the phenomenon of hysteresis where constantly magnetising and demagnetising a conductor also creates heat. One of the first researchers to explore the numbers behind magnetic hysteresis was Ernest Rutherford.

INVERSE SQUARE LAW

The inverse square law describes the way radiated energy spreads out with distance. Because the area expands exponentially, while distance is linear, this means that the energy delivered to any given area drops dramatically with increasing distance.

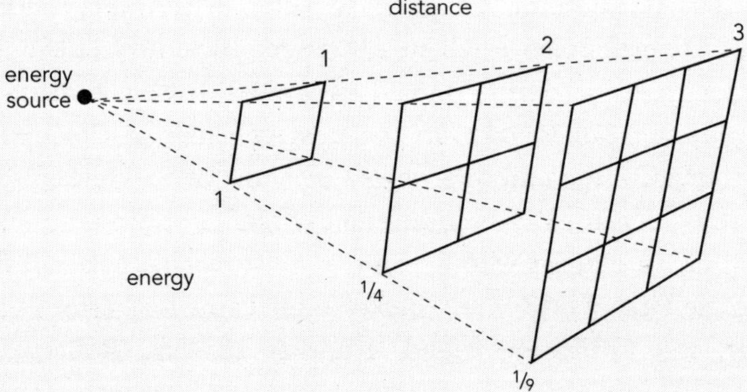

(1865–1941), later felt that Bickerton's 'highly original' techniques for teaching suited Rutherford.[20]

Christchurch was the main city in the South Island, a regional focus of economic power and a bustling metropolis that had emerged from the ambitions of the Canterbury Association. This had been set up in 1848 to establish a million-acre (404,000 hectare) Anglican church community in New Zealand, again on the basis of Wakefield's 'instant city' concept.[21] It was laid out in the square-block fashion of the day, becoming a municipal district in 1862. In a deliberate effort to echo the magisterial feel of British metropolitan centres, major public structures were built in stone, notably the massive Christ Church Cathedral, around which the city layout pivoted. A museum building was begun in 1870 just a few blocks from the cathedral; and then in 1877 a new stone building was completed for Canterbury College on a site just opposite the museum.

Christchurch introduced Rutherford to metropolitan life. By 1890 the city had all the latest conveniences. Gas streetlights were introduced in 1862 along with water services. Work on a municipal sewer system began in 1879, just four years after London's own modern systems were completed. Steam-driven trams began working central city streets in 1880, and by 1890 efforts were under way to bring in electricity. Christchurch, in short, was a modern, go-ahead centre, the largest city in the South Island, increasingly prosperous on the back of a rising economy.

Rutherford enrolled for a Bachelor of Arts, a three-year course built around Latin and pure mathematics. He enrolled for French and applied mathematics, adding physics as an extra. This was not just a figurative extension. Bickerton taught in a makeshift laboratory in a corrugated iron building nearby, nicknamed the 'tin shed'.[22] This mix of the sciences with an arts degree was not unusual for the day. At age 19, Rutherford was described as an unexceptional student, 'boyish [,] frank, simple and very likeable ... with no precocious genius'.[23] He got his science from Bickerton, but mathematics came from Professor C.H.H. Cook, a 'very orthodox and strictly scholastic' teacher who 'acted as a brake to the young student and prevented his speculations from leaving the ground of fact.'[24] On the strength of Cook's teaching, Rutherford won First Class honours, giving him the Mathematics Exhibition[25] that year — a prize of £20.

Rutherford also joined the College Rugby Club.[26] In some ways it was an odd choice: he had played rugby at high school in a manner later described as 'without distinction.'[27] He continued to play in the Canterbury team through his university years. Inevitably, the fact of his playing a game that by 1890 was on the way to becoming New Zealand's national sport became entwined with mythologies swirling about his life.[28] The idea that this world-leading scientist was also an ordinary rugby-playing Kiwi struck chords with New Zealand's twentieth century 'blokish' self-identity as a land of 'rugby, racing and beer', reducing Rutherford to normality and making him 'one of us' to everyday New Zealanders. This vision spoke more about New Zealand society than it did about Rutherford. He was generally unprepossessing. William Marris, his fellow student, recalled Rutherford as 'very modest, friendly but rather shy and rather vague', someone who had 'not yet found himself.'[29] This reservation perhaps explains Rutherford's profile with the various student associations he joined, notably the Science Society, where he attended meetings but seldom spoke.

In his second year Rutherford enrolled in Bickerton's course in practical physics. Although records are unclear it is likely that this was the year when he found private accommodation in a boarding house owned by 39-year-old Mary Newton. She had four children, had been widowed in 1888, and was making ends meet by offering lodging to senior college students. One of the outcomes was that Rutherford struck up a friendship with Mary's eldest daughter, Mary Georgina Newton — May to friends and family, who was attending high school.[30]

Bickerton's course drew Rutherford into hands-on electrical apparatus, an array of Gieseler tubes — the cousins of the Crookes tubes that Rutherford later used at Cambridge — and galvanometers and induction coils able to produce high voltages.[31] These became central to his subsequent work. In August, Rutherford helped demonstrate this equipment to the public during a conversazione, a public exhibition of the university's science gear. It was organised by the Canterbury College Science Society, which had been formed only a few months earlier.[32] Between 500 and 600 people attended according to one newspaper,[33] and all, according to another, 'seemed to enjoy themselves.'[34] Rutherford wrote home next day to explain that he was 'boss' of the 'darkroom', a space in which he had set up high-voltage equipment, tricky to assemble in the dark.[35]

The following year, 1892, was Rutherford's last as an undergraduate and his last under his original scholarship. Senior scholarships were available by external examination, and at the end of the year Rutherford sat exams in mathematics and experimental science. That enabled him to return to Canterbury for a one-year Master's degree, making him one of only 14 post-graduate students in New Zealand.[36] He hoped to get a double first-class pass in physics and mathematics, becoming the only Masters-level maths student in New Zealand at the time. He also enrolled in chemistry, chemical laboratory practice, biology and botany.[37] The course included a research thesis. Rutherford decided to look into the way iron was magnetised by alternating current.

Rutherford's focus in 1893 was driven by recent developments in alternating current. An iron core inside a copper winding was known to improve the efficiency of the transformers that converted AC into DC. But nobody knew whether metals could be magnetised and demagnetised at very high rates. Nobody knew exactly how much energy would end up being dissipated into the iron as heat. Was such dissipation related to the frequency of the current? What was the rate of magnetic penetration? This was not idle curiosity. Practical transformer designs, among other applications, needed hard data. That relied on detailed investigation into every aspect of the field, each small discovery adding detail to a growing picture.

His approach was straightforward: he used simple but ingenious equipment to produce a remarkably thorough array of empirical data. This approach — not merely exploring the unknown, but actively measuring it in detail — became a hallmark of Rutherford's work. The techniques he used evolved over time, but all reflected the style he devised at Canterbury University. This came about in part by necessity. When he began his first research thesis at Canterbury he discovered the British physicist and head of the Cavendish Laboratory, J.J. Thomson, had been running similar tests but had not quantified the results.[38] Rutherford decided to provide numbers and tested materials ranging from soft iron to carbon, copper, platinum and steel filings. He tested various thicknesses, including steel sewing needles and piano wire, scribbling the results down in his notebook.[39] He dissolved some of his test objects in dilute nitric acid to measure when they stopped responding to the electromagnetic field and, thus, lost their magnetic properties. This enabled him to identify the depth at which the material was magnetised. He used Leyden jars to power his tests, achieving currents of up to 100 amperes, a dozen times the maximum draw of a modern portable fan heater.[40] Rutherford spent six months on this work, which he initially called 'Magnetism of iron by Leyden jar discharge', drafting his first paper in red and black ink in his former biology notebook.[41]

By this time Rutherford had reached the limits of the New Zealand system. The only option now was to head overseas — which for Rutherford meant either Britain or Germany, the two key centres of world physics. The problem, as always, was paying for it. Rutherford discovered he could get funding for overseas study through an 'Exhibition of 1851 Scholarship'. This had emerged from a major trade exhibition held that year, initiated by Queen Victoria's husband Prince Albert to show off British technology. It was a roaring success, producing a profit of £186,000, equivalent to about £21.8 million in mid-2020s values.[42] The Royal Commission behind the exhibition was granted a charter to use the money for scientific and educational purposes, and initially it went toward museums, but in 1891 the Commission established a scholarship designed to provide £150 annually for two years to students from Britain's far-flung colonies.

New Zealand was offered one entry for 1894. The University of New Zealand asked to defer selecting nominees until 1895, a decision that made it possible for Rutherford to prepare a paper during 1894.

Meanwhile Rutherford finished his Master's degree, and at the end of the 1893 returned to Taranaki to wait for his exam results. He applied for work at Christchurch Boys' High School as a teacher, one of the few careers open to someone with his extensive list of qualifications. In the event he did not get the job, and returned to Christchurch for the 1894 academic year because he could only apply for an 1851 Scholarship if he was a student. This time Rutherford enrolled in science courses to build credits for a Bachelor of Science degree. Soon afterwards he received his 1893 exam results: he had achieved his Master of Arts with double First Class honours.[43]

Behind it all loomed the 1851 Exhibition Scholarship. Rutherford needed a body of research to back an application and decided to explore the rate at which magnetisation travelled through metal, known as its 'magnetic viscosity'.[44] This was new ground. At the time metals were known to be made of atoms. These were thought to be solid, and the rate at which they could be magnetised was thought to be a product of the rate at which these atoms moved. However, specific figures had never been produced. Rutherford was about to begin work when he received a copy of the April 1893 *Proceedings* of the Royal Society in London, which — as he put it — 'contained an account of experiments by Messrs Hopkinson, Wilkinson and Lydall, which in a great measure anticipated what I had intended doing.'[45]

This did not deter him, as he felt he could deal with shorter time periods than they had been able to measure. To do this he built an electro-mechanical instrument able to measure intervals down to a millisecond via an ingenious system in which a weight fell on to two levers that were slightly out of alignment, thus switching on a current at slightly different times.[46] Rutherford could precisely adjust the misalignment by means of a calibrated screw, and the difference told him the time interval.[47] This apparatus captured the essence of Rutherford's approach to all physics problems — one he applied time and again through his lifetime. He was a mathematician first, and his equipment design flowed from his analysis of the problem, blending ideas that ran from classical Greek trigonometry to Newton's gravitation.[48] Using an electrical circuit devised by Hertz, Rutherford was also able to produce alternating current frequencies of up to 100,000,000 cycles a second. This was within the range of what was later used for domestic radio — in fact a Christchurch FM station, Newstalk ZB, was broadcasting close to that same frequency 130 years later.[49]

One result was that Rutherford found a way of using steel wires to detect oscillating electromagnetic fields. According to his fellow student Robert Laing, Rutherford's equipment was able to pick up fields 'from one end of the old "tin shed" to the other'.[50] This was about 20 metres and does not sound like much by twenty-first century standards — such a range is typical of any home Wi-Fi setup — but in 1894 this was extraordinary.[51] Detection at range was not the main aim of Rutherford's researches at the time; but it meant that he had a form of radio detector. Such a device was the focus of deliberate investigation elsewhere. British physicist Sir Oliver Lodge produced an adaptation of Branly's detector the same year at University College in Liverpool. Efforts a little later by British naval officer Captain Henry Jackson and the Italian inventor Guglielmo Marconi led to workable radio.[52] Rutherford's detector had a specific feature: the electromagnetic field was picked up only one way.[53]

Rutherford produced a second paper from his 1894 experiments,[54] which he submitted to the University of New Zealand. This got him a nomination for the 1851 scholarship. The other nominee that year was a chemist, James Maclaurin (1864–1939), who produced a thesis analysing the effects of potassium cyanide on gold.[55] Both papers were then sent to the 1851 Commission in London for assessment. MacLaurin won. It was a finely balanced decision: there was an argument within the Commission over whether to offer an out-of-cycle scholarship to New Zealand that Rutherford might apply for the following year. But then Maclaurin declined the award: he was getting married and had decided to settle in Auckland.[56] The scholarship defaulted to Rutherford. The news reached him in July when he was at his parents' property in Taranaki. He later insisted that he was digging potatoes when his mother arrived with the news, so he threw down the shovel and declared he was finished with such things.[57] It was a good story, but there is no indication that it is true: Rutherford had aspirations but did not know what his future held. The imagery, again, leans into his self-image of being from humble farming stock.

More to the point is the fact that he was second choice. It was a curious echo of his earlier quest for scholarships for high school and university and his applications for work. Each time he only just squeezed in, variously on the second attempt or through circumstance. The consistency tells us much about Rutherford. He had trouble communicating clearly on the spot, meaning he struggled to perform well in exams and to sell himself to potential employers. His scientific work — pursuing what at the time presented as niche interests — always seemed to just miss what was wanted. He got there in the end, but his early experiences make clear that it was in part by chance. A dogged determination to try again after each setback got him through. But even so, the difference between Rutherford the world-leading scientist and Rutherford the drop-out seems slender.

5 | A good keen scientist

Mr Rutherford possesses, to a marked extent, that valuable quality which I can best indicate by the phrase scientific keenness.

— E.H. Griffiths, Fellow of Sidney Sussex College, Cambridge, May 1898[1]

THERE IS CIRCUMSTANTIAL EVIDENCE that Rutherford never expected to get an 1851 Exhibition Scholarship. When it arrived he had done nothing to pave the way for overseas study and had only a general idea as to what he might do. He anticipated finding opportunities at Cambridge or — if that failed — Berlin.[2]

Now, scholarship in hand, he decided to make arrangements once he got to London, and he moved with remarkable speed. The telegram advising his award reached the University of New Zealand on 9 July.[3] Rutherford likely received it the next day, and was off, whipping together the documents he needed — including his birth certificate, with its spelling mistake — and organising his journey around the world. He booked passage on the Union Steam Ship company's steamer *Wakatipu* for Sydney, intending to then catch the P&O liner *Himalaya* for England. He borrowed money to pay for the tickets and left New Plymouth by coaster on 22 July, arriving in Christchurch early on the 25th.[4] Then on 31 July he boarded the *Wakatipu* at Lyttelton. Mary travelled with him as far as Wellington.[5]

Rutherford departed Wellington for Sydney on 1 August, just three weeks or so after he discovered he had the scholarship. It was his first time out of New Zealand, and we can imagine his sorrow at leaving Mary, mixed with his excitement at the unknowns ahead of him. In a historical sense he was one of the first New Zealanders to engage in what later became known as the OE — 'overseas experience', a rite of passage and validation of worth for generations of twentieth-century Kiwis that pivoted around Britain being 'home', even for those who had never been there.

Rutherford's go-getting nature when it came to science became obvious when the ship reached Sydney and he discovered the *Himalaya* had departed the day before for Melbourne. He booked a train to catch up with it there, and meanwhile had time to introduce himself to Sydney University. He reached Melbourne and joined the ship, then took advantage of a two-day stop in Adelaide to see William Bragg, a British physicist also working in electromagnetics. Here he showed Bragg his electromagnetic detector.[6]

Rutherford's choice of London as the starting point for his adventure was no coincidence. The Imperial capital of 1895 was the place of dreams for colonial New Zealanders, the essence of 'home', its landmarks names to conjure with. Ironically, the city didn't suit Rutherford: he arrived in late September and promptly got a sore throat, then began suffering attacks of nerve pain.[7] He collected the first instalment of his scholarship money from the University of New Zealand's agent in London, then wrote to Thomson asking whether he could work in the Cavendish Laboratory, enclosing a paper on his research in New Zealand.

Thomson did not wait to read the paper and wrote back wanting a meeting.[8] That Friday, 27 September, Rutherford took an express train to Cambridge. He met Thomson, reporting to Mary that he found the great man 'very pleasant in conversation ... not fossilised at all.'[9] They talked in the laboratory for a while, then Thomson invited Rutherford to dinner back at his home on Scroope Terrace, where Rutherford met Thomson's wife Rose and their young son George. Thomson was keen to have Rutherford on board. The university had just allowed overseas students to join and Thomson suggested that Rutherford do so.[10]

Everything had fallen into place, and surprisingly easily. Rutherford returned to London, where his neuralgia flared again. He packed up his belongings, and the following Tuesday, 1 October 1895, arrived in Cambridge.[11] Rose Thomson found him a place to live — 'diggings', in student slang of the day. As in Christchurch, the university town had a brisk market in rooms to let, and here Rutherford had his first shock. London had been expensive, but that was the Imperial capital. Now he found that Cambridge accommodation was also costly, and his scholarship could only stretch so far.

The world Rutherford found at Cambridge was hauntingly familiar in some ways. The stone architecture of Canterbury University had been intentionally designed to match that of British universities. In other ways, though, life was different. For all its conceit of being Britain's 'best child' and its adulation of British ways, New Zealand's late-colonial culture was distinct from that of its parent. The social shifts that went with industrialisation had done much to even out long-standing British cultural strata. But that stratification was still there, in contrast to the New Zealand experience where there were economic distinctions, but the primary frameworks remained the socially flattened middle class ideals of the original settlers.[12] Into this social stratification were mixed innumerable small differences, right down to slang, and that was without considering the subcultures of academia. Rutherford discovered these when he arrived at Cambridge. The university had been founded in 1209, pre-dating the entire human history of New Zealand.[13] The institution was steeped in tradition, prestigious, and one of the top universities in the British Empire.

The remarkable part was that Rutherford arrived there in one bound from New Zealand, engaging physics directly in the main British research institution of the day, after study at a relatively minor colonial university. His publications and remarkable array of degree qualifications helped, but even so, it is unlikely such a leap could have happened so easily in later years. Part of the reason was scale. Science was institutionalising across the Western world during the late nineteenth century, but in the 1890s that scale was modest by later standards. Rutherford could go directly to Thomson without being intercepted by a staff member delegated the task of handling student enquiries. Thomson, in turn, had time to respond promptly. This same phenomenon also explains why the physics of the age could be dominated by a relatively few individuals, albeit now backed by teams of students and colleagues.

Into that flowed a healthy dose of luck, including MacLaurin's decision to pull out of the 1894 scholarship. Luck was also on Rutherford's side at Cambridge. The university had just opened its doors to external research students. Rutherford had not gone there to become a student but now followed Thomson's advice and, with the Irish student John Townsend, became one of the first external research students at the university. The term was due to begin on 10 October, and Thomson urged Rutherford to join one of the colleges. He chose Trinity, joining it in a brief ceremony with about 200 others on matriculation day, 20 October. It cost him £20 of his scholarship, of which £15 was returnable for good behaviour.[14] This was no small sum in period terms.

Thomson found that Rutherford was go-getting and had brought 'his own problem with him, which he had begun in N.Z. [his] instrument for detecting wireless waves'.[15] And so Rutherford picked up where he had left off in New Zealand, exploring the potential of his electromagnetic detector. Radio was the hot topic of the year, and not just in the rarefied air of physics laboratories. The Royal Navy was interested. Communicating with ships at sea was becoming a priority. Commercial shipping companies were also quick to see the potential.

Theory and technology came together during the early 1890s, and by the middle of that decade the question was not whether wireless telegraphy could be achieved, but when and where. Multiple efforts were under way, notably in Britain. The Italian inventor Guglielmo Marconi produced a device able to transmit over a short distance in 1894, and patented it. By 1895 he was working in Britain to produce a practical apparatus.

Rutherford's work at Cambridge paralleled that of Marconi and others. He did not work alone: his gear had to be carried around and needed someone at the other end to confirm reception. By early 1896 he was pushing for distance, writing to Mary that he hoped to detect a signal from the tower of St John's, a little under a kilometre distant.[16] On 28 February he roped in Townsend and another student, John McClelland, to run an ambitious test between Townsend's rooms and the Cambridge common. He set up his receiver in Townsend's rooms around 6.30 p.m. The device included his detector, a cluster of steel needles set inside a coil. The system was straightforward: when a vibrating electromagnetic field swept over the instrument the coil demagnetised the needles, causing a small mirror to swing.[17]

Nothing went well. The first problem was creating a field to detect. Rutherford's transmitter consisted of two parallel steel plates carrying conducting rods ending in brass knobs, along with the batteries, coils and other equipment he needed to generate high-voltage AC. The gear was cumbersome, and that evening Rutherford arranged for two men with a handcart to wheel it to the common, where Townsend and McClelland could set it going. The labourers mistook the agreed time and had gone before Rutherford arrived. He and his colleagues then asked two passersby to help, and by around 9.15 p.m. the transmitter was on its way to a site around 500 metres from Townsend's rooms. Rutherford couldn't detect a signal. They reduced the distance. Still nothing. Finally they got the range down to 100 metres, which had worked before. Nothing. Rutherford then tried with a different detector and, finding a signal at 350 metres, realised his main detector had developed a fault. They returned the equipment to the laboratory around 11.30 p.m. It had been an exhausting night.[18]

From the perspective of twenty-first century physics research, with its complex and expensive equipment operated by large institutions, the nature of these experiments seems almost slapdash, an evening jaunt by students. And yet it was at the leading edge of radio as it stood in 1896, paralleling experiments by Marconi, Lodge and others. Rutherford was well aware of the potential, writing to Mary that if he could make it work at 10 miles (16 km) he could make money out of it.[19] Rutherford was shortly invited to join Thomson's work on X-rays and particles, but gave a presentation about his radio detector to the Royal Society in June 1896, and another in September to the British Association in Liverpool.[20]

However, Rutherford's focus increasingly became aligned with Thomson's. In this way Rutherford swung into the next leading edge of physics, working with Thomson to discover ways that X-rays — which appeared to be an electromagnetic phenomenon — could be used to further explore the nature of matter. He settled into Cambridge life, joining his fellow students on a bicycle holiday to Lowestoft. He also bent to peer pressure to start smoking a pipe. He thought it might help him relax while working in the laboratory, writing to Mary to explain that scientists 'ought to smoke.'[21]

To give Rutherford's work in Cambridge context we need to turn to some of the other developments in electrodynamics of the period — discoveries that shaped Rutherford's career. One burning issue remained the aether. By period thinking it had to exist, otherwise electromagnetic energy could not propagate. But nobody could detect it. Michelson wondered if the aether was being fully dragged along with the Earth and thus in effect stationary. This could be one explanation of his failure to discover differential drag effects, but this idea was neither testable nor compelling.

Others had different opinions. Oliver Heaviside (1850–1925) reworked Maxwell's electromagnetic equations and demonstrated in 1888 that electromagnetic fields changed shape when they moved:

the force lines compressed in the direction of that movement. The following year the Irish physicist George FitzGerald wondered whether this same phenomenon applied to solid matter. This was superficially weird, but it was not implausible: all that was required — as the German physicist Max Born observed later — was to reverse the order of priority. Instead of Euclid's geometry and Newton's mechanics framing electrodynamics, perhaps electrodynamics framed geometry and mechanics?[22]

This shift of concept was an important step in the evolution of electrodynamics as a driver of physics, opening the door for Albert Einstein's later work. But in the late 1880s the key appeal of FitzGerald's idea was that it explained Michelson's failure. By this argument, Michelson's equipment had contracted in the direction of movement in such a way as to compensate for any aether drag. Naturally this was undetectable, because everything contracted the same way. It also turned out that FitzGerald was not the only physicist thinking along these lines. The Dutch physicist Hendrik Lorentz (1853–1928) came to the same conclusion in 1892, adding a factor he called 'local time'. What became known as 'Lorentz-FitzGerald contraction' put the aether back in play. But it was not a complete explanation, nor could it be tested.

In hindsight the value of FitzGerald's idea was the way it questioned classical assumptions. And Einstein later picked the idea up and ran with it, showing that FitzGerald was right. The irony is that this idea was raised not to create a new framework for understanding physics, but as yet another effort to make observed reality fit the theory of aether. This meant in practice that by the early 1890s the notion of an invisible aether was becoming tangled with complex and often counterintuitive add-ons trying to keep the concept alive. Most had enough evidence behind them — in terms of the science then known — to appear credible. But none were testable and most were inconsistent with each other. That was worrisome. Yet without the aether there could be no electromagnetic transmission unless physicists had missed something.

The only answer was more research, as physicists patiently assembled the pieces of the electrodynamic jigsaw and explored how that related to the structure of matter. It was a painstaking process in which new discoveries often raised more questions than they answered. One key question in the mid-1890s was the nature of atoms. Thomson, as we have seen, was able to identify the first subatomic particle by 1897 and show that it carried an electric charge. That raised other questions. Were there more particles? Where did they sit in terms of electromagnetics?

Into that collection of questions rushed a new mystery: the behaviour of uranium, which unfolded to the scientific world partly by accident. Uranium itself was well known by the 1890s, typically as an additive to glassware. Uranium salt crystals were known to glow when exposed to sunlight, and in the mid-1890s the French chemist Henri Becquerel (1852–1908) decided to wrap that in proper science. He speculated that sunlight was being absorbed and re-emitted, which was a reasonable proposition at the time: the principles of thermodynamics made clear that energy had to be conserved. Röntgen's discovery of X-rays threw a new question into that work, to which Becquerel was alerted in January 1896 by the French mathematician Henri Poincaré (1854–1912).

Late in February Becquerel began a series of experiments to test whether the uranium fluorescence included X-rays. His test rig was simple: he sandwiched a coin between crystals of potassium uranyl sulphate and a carefully wrapped photographic plate. He then exposed this setup to the sun for several hours. The result when he unwrapped the plate and developed it apparently confirmed his suspicion: the wrapped plate had received no light, but it carried a shadowy image of the coin. He wanted to repeat the experiment on 26 February, but clouds rolled in. Paris remained stubbornly overcast next day. Dismayed, Becquerel put the photographic plates away in a drawer and left the crystals on his desk above it.

Becquerel did not touch the crystals the following day, 28 February. It was a busy day for science, though: that evening, on the other side of the Channel, Rutherford and his colleagues were trying to get results from his radio detector in Cambridge. It was a leap year, and Becquerel did not move the crystals on 29 February. Then, on 1 March, he decided to develop the plates he had put away in the drawer. To his surprise they were heavily fogged. This was extraordinary. The shadow image he had obtained three days earlier, it seemed, had not been due to fluorescence driven by absorbed sunlight. He tested that idea and found it was true — indeed, by 9 March he had kept uranium crystals in the dark for 160 hours without any reduction of their invisible emissions.[23]

This was an entirely new discovery. Some kind of energy was being emitted from the salts, with effects on photographic plates that were similar to the X-rays Röntgen had produced electrically with a

URANIUM — IT'S ALL IN A NAME

URANIUM WAS FIRST IDENTIFIED in 1789 by the German chemist Martin Klaproth (1743–1817) who found it in an ore then known as pitchblende, now known to be primarily made of uranium dioxide and its decay products. The name uranium came out of an ongoing debate over what to call the newest planet, which William Herschel discovered in 1781. Herschel called it Georgium Sidus (George's Star) after the British monarch, George III.[24] That did not go down well in Europe. Various alternatives were proposed, and German astronomer Johann Bode suggested the classical Greek god Ouranos, spelling it in the Latin form, Uranus. It was an odd choice. Jupiter and Saturn had been named after Roman gods, which should have made the new planet Caelus, their equivalent of Ouranos. The general debate over what to call the new planet was far from settled in 1789, so Klaproth reputedly decided to name the new element uranium in support of Bode's proposal. Klaproth's new element was finally isolated in 1841 by the French chemist Eugène-Melchior Péligot (1811–90).

Crookes tube a few months earlier. The problem was that this uranium emission did not require any input, which defied the well-established laws of energy conservation. A mystery. Becquerel put much of the rest of the year into exploring the phenomenon. In other years this might have thrown him into the limelight, but this was the year of rays — X-rays, cathode rays, radio transmission and more — with the result that Becquerel's new rays took their place alongside the rest in a science community awash with new data. It was another piece in a jigsaw puzzle where physicists had neither the final picture, nor yet any idea as to how many pieces they needed to construct it.

The full import of Becquerel's discovery had to wait for two new players, Marie Curie (1867–1934) and her husband Pierre (1859–1906). Marie was born Maria Salomea Skłodowska in Warsaw, youngest of five. Her father was a mathematics and physics teacher and coached Maria, who showed a deep talent for mathematics. She was educated to secondary level, briefly becoming a teacher. Some of her salary went to fund her older sister Bronia's medical studies in Paris. In 1891, Maria went to Paris to study physics and mathematics at the Sorbonne. She Gallicised her first name.[25] It was an unusual career to choose: women were scarcely seen at higher levels in any of the sciences and — at the time — were barred from physically entering some of the institutions.

Marie arrived in Paris at the height of what was later known as France's *Belle Epoque* — 'Beautiful Era' — a period of bold exploration of new culture, art and science buoyed in large part by Imperial power. The name emerged later amidst a sense of nostalgia and loss after this age was broken by the First World War, but it was appropriate. Paris stood at the centre of it all, home to impressionist artists who entwined light and mood on canvas; to the composer Claude Debussy whose ambition was to write music that made listeners imagine colours. It was home to poets, writers and thinkers, a city of light, of bohemian excesses and of morning-after ennui. The arts flourished, and so did the new sciences. As a student Marie had little money, but pressed on despite the challenges of poverty and of being a woman in a male-dominated world. She worked in the laboratory of Gabriel Lippmann during 1894 and met Pierre Curie. He was completing his doctoral studies under Lippmann on magnetism, discovering that ferromagnetic materials lost that property above certain temperatures. Pierre and Marie married in July 1895.

To Marie, Becquerel's discovery posed questions, and she began investigating. Pierre stepped in to support her work. Like Rutherford they used home-baked instruments, including a piezoelectric device developed by Pierre with help from his brother Jacques that produced voltages when under physical pressure. This enabled Marie to discover whether what she called 'uranic rays' could ionise gases. It was difficult work, not eased by the fact that — as Marie recalled later — the School of Physics accommodated them in an old shed that had once been the dissecting room for the School of Medicine.[26] There were few results until February 1898 when Marie discovered that both pitchblende and thorium compounds produced the new rays. The interesting part was that pitchblende was known to contain uranium, but was producing more energy than her experiments had shown could be created by uranium alone. In April, she and Pierre published a paper with their initial conclusions, following that with a paper that revealed a new discovery, an element isolated from pitchblende that they dubbed polonium, in honour of Marie's native country. This paper also introduced a new compound word to science, 'radio-activity' — originally with the hyphen — that Marie coined via Latin to describe the fact that these elements radiated energy. Later, as so often happens to English compound words, the hyphen was dropped to create the neologism 'radioactivity'.

Then in December the Curies revealed another element — also extracted from pitchblende — that Marie dubbed 'radium', again drawing from Latin. This was highly active and largely explained why pitchblende emitted more radiation than if it contained uranium alone. At this point they had only enough radium to show that it existed. More information, such as its atomic weight, needed larger samples. The problem was that radium was present only in trace quantities through the ore. So the Curies set about extracting more, a laborious and time-consuming task, but necessary to learn more about this new material. From a modern perspective the idea of these two scientists trawling through relatively enormous quantities of radioactive ore to pull out its most dangerous component seems foolhardy. However, the Curies were pioneering a new field. By this time X-rays were known to produce burns after moderate exposure. Other effects were not yet realised. But the Curies continued their work even when these risks were known.

These discoveries were ground-breaking. A new property of matter had been found, wholly unknown in classical times and which was entangled with the new world of rays. In 1903, Becquerel and the Curies shared the Nobel Prize in physics for their joint discovery of the phenomenon.[27] The fact that this was awarded in physics

THE DARK SIDE OF RADIATION

MARIE CURIE'S WORK with radioactive elements continued for many years, even after the dangers became better understood. One result was that her papers, including her cookbooks, became so badly contaminated with radium-226 that later they could not be handled without safety equipment. Even letters she despatched to Rutherford by ordinary post were sufficiently radioactive to be tagged hazardous by the Cambridge University Library, requiring special precautions when they were researched for this book.[28]

The tragic result was that, despite taking precautions once the dangers were understood, Marie became increasingly ill from radiation exposure and died of aplastic anaemia in 1934. At the time her body was considered to be sufficiently radioactive that she was buried in a lead-lined coffin, something then forgotten until 1995 when she and Pierre were exhumed for reburial in the French national mausoleum.[29] Her daughter Irène and son-in-law Pierre Joliot, both of whom also worked on radioactive materials, also died of radiation effects. Pierre Curie might also have suffered this fate, but was killed in a coach accident in 1906.

was curious: their work largely involved hands-on chemistry. Later, Rutherford received the Nobel Prize in chemistry for his investigations of radioactivity that, broadly, involved hands-on physics. The issue underscored the fact that these divisions were somewhat arbitrary, a projection of the way Western society of the day saw the world. In the late nineteenth century, with the discovery of phenomena that spanned these human-constructed divisions, the conceit broke down.

The new elements found by Becquerel and the Curies raised questions for physicists and chemists alike. How could an element produce energy, apparently out of nowhere? And how did that relate to what was being discovered about the relationships between electromagnetic energy and particles?

This became the focus of Rutherford's work from the late 1890s, and it might be easy to suppose that he was somehow destined to dominate this new world. In fact, as we have seen, Rutherford's career to this point was unstructured. He had gone to university in part because he did not know what else to do. Here his enthusiasms had been seized by electrodynamics. By the time he graduated he saw a future in the academic world, but at this point did not know what that might be. His focus was on whatever his latest fascination had become. When he took his electromagnetic detector to Cambridge he first extended that to radio, then veered into the work suggested by Thomson, which swiftly captured his imagination at the expense of his previous focus. It also offered direction. His work with Thomson on X-rays during the summer of 1896 led to a joint paper that Thomson and Rutherford presented to the British Association, subsequently publishing in the *Philosophical Magazine* that November.[30] For Rutherford this was a major step up, attaching his name to one of the most prestigious physicists in the field. It reinforced his place as Thomson's star student. And it also reinforced his new direction.

As the year waned Thomson put Rutherford to work on his own, analysing the interaction between X-rays and various trace gases. Rutherford was fascinated. 'Don't be surprised,' he wrote to Mary in October, 'if you see a cable some morning that yours truly has discovered half-a-dozen new elements, for that is the direction my work is taking.'[31] Just two years earlier Lord Rayleigh and William Ramsay had discovered argon by analysing the molecular weights of nitrogen produced by two different means. The electrical measurements Rutherford was making of ion decay timing revealed differences between gases, and we may speculate that Rutherford thought he might find a new element or two hidden within the gases he was studying. But he admitted to Mary that the probability was low.[32]

Rutherford pushed hard, working seven hours daily in the lab, coaching students and reading, then planning research from the quiet of his rooms during the evening. The outcome was a raft of data that put numbers to a phenomenon later harnessed for fluorescent lighting and neon signs.[33] In November, Thomson set Rutherford to work on hydrogen ions.[34] Rutherford dived in, his work revealed in the notebook pages for these months.[35] The techniques he applied were broadly those he used for the rest of his life, exploiting his understanding of electromagnetic phenomena to produce detailed measurements. Once again his focus became the intersection between gadgetry of his own invention, mathematics, and the information this equipment could produce.

The data Rutherford obtained — thorough, detailed and with the parameters well thought out mathematically — supported Thomson's own research. Early in 1897, Thomson concluded that the charge on the X-rays produced by a Crookes tube pointed to a particle around a thousandth the mass of an atom. It was the first subatomic particle to be discovered. As we saw in chapter 1, Thomson called these 'corpuscles', and it took a while for the name 'electron' to settle down. But with this announcement it could fairly be said that the field of subatomic particle physics was born.

Rutherford spent much of his time with colleagues, living the academic life in Cambridge. But he made room in his schedule when Mary arrived in the northern spring of 1897. Rutherford's conservative attitudes framed the moment. They were engaged but not married, so she stayed for a week with the Thomsons while Rutherford remained in his rented rooms. They watched boat races on the Cam, and later went to Ireland and stayed with friends. It was during Mary's visit that Rutherford discovered something else that soon seized his interest. They went to the Crystal Palace where he had the opportunity to see one of the first motor carriages. He thought they might be cheaper to run than a horse.[36] The incident triggered a life-long enthusiasm for cars.

The interlude was brief. Mary returned to New Zealand, still engaged but without a specific date for their nuptials, largely because Rutherford was worried about how to fund married life. He turned back to his work. His imagination had been seized by the tasks Thomson had set him, and after exploring X-rays he switched to ultraviolet, the part of the electromagnetic spectrum just beyond visible light. Hertz had discovered that this wavelength interacted with electrical discharge — the photoelectric effect — and Rutherford wondered whether it could produce ions. If it did, were these the same as the ones created by X-rays?

As Rutherford worked on these questions the discoveries across the Channel at the hands of Becquerel and the Curies drew his attention. Early in 1898, he wondered whether 'uranium rays' could also produce ions, and whether they were similar to the ions he had produced with X-rays and ultraviolet light. There was only one way to find out, and he turned to the task as soon as he had finished with ultraviolet. He was particularly interested in finding out whether he could polarise these 'uranium rays'. Becquerel had run experiments suggesting that was possible.[37]

What followed was a remarkable and thorough-going effort to examine the radiation emitted by both uranium and thorium from every aspect: its polarisation, whether it ionised gases, its complexity, and the degree to which it was absorbed. This characterised Rutherford's approach: he didn't ask just one question. He asked as many as he could think of. There was also one major difference between Rutherford's work and that of the French physicists. The

French had been using photographic plates to detect uranium 'rays', but that took up to a week for each experiment. Rutherford used photographic plates for his polarisation experiments, but decided he could use electrical methods for ionisation tests, giving instant results and, more crucially, numbers. To do this he added an electroscope to the system he already had for detecting ions via a galvanoscope.

Electroscopes had been long used to identify static charges and consisted of a thin piece of aluminium foil attached by one edge to an electrified plate. The angle at which the foil rose from the plate determined the magnitude of the charge, and the gradual change in that angle showed how the charge leaked away. Rutherford set this up in close proximity to a second plate with opposite charge, in such a way that ions created by uranium 'rays' could be blown between two plates. The ions caused the charge from one plate to leak into the other, which he could measure with the electroscope. It was an ingenious system that meant he could obtain quantifiable results quickly. He could also test various materials to see if anything stopped the uranium 'rays'. Rutherford then moved on to test whether uranium radiations were absorbed by gases, using a new and electrically complex piece of gas-tight apparatus, which he ended up coating in tinfoil to prevent static charges building up and ruining the results.[38]

It was a thorough effort to fully examine the radiation from uranium, and from this Rutherford discovered two different types of rays. One was able to be stopped by a thick sheet of paper. Rutherford drew from the Greek alphabet and called this the 'alpha' ray. The other was more energetic and would penetrate further. He dubbed this the 'beta' ray. This was the first identification of what — thanks to his further work — we now know to be alpha and beta particles. The alpha particle, in particular, became pivotal to Rutherford's subsequent career, first as the focus of his investigations and then primarily as a research tool. The associated terms, alpha and beta radiation, remain in use today.

As always, Rutherford found one discovery leading to more questions, pushing him to explore further. He extended his work to thorium, another element that had been found to produce 'rays'. This was first identified in its mineralised form by the Swedish chemist Jöns Jacob Berzelius in the late 1820s. He called the mineral thorite after the Norse god of thunder. Nineteenth century chemists did not yet have the technology to extract the pure metal, instead obtaining various salts and oxides. It did not take long to discover that thorium oxide glowed when subjected to the heat of burning gas, without itself being consumed. This led to it being used for gas mantles from the mid-1880s. The fact that thorium-based mantles were also radioactive was not known until the Curies discovered the point later. The nature of that radioactivity — which produced radon-226 — was discovered later still by Rutherford. None of this prevented ongoing use of thorium oxide in portable gas lights.

Rutherford found thorium emissions to be more penetrating than uranium. However, he could not get consistent results — and by this time other priorities were intruding. He was relying on his 1851 Exhibition Scholarship money, and it would not last forever. His growing prestige helped convince the trustees to renew the scholarship for 1898. He was also able to glean some money from tuition, partly paid for by Thomson. Late in 1897 he applied for the Coutts-Trotter Scholarship offered by Cambridge, and received it. This gave him £250 annually in addition to the £150 Exhibition money. 'Think of it,' he wrote to Mary, 'nearly enough to get married on.'[39] In practice it probably was sufficient: the two scholarships together totalled just over £43,500 in mid-2020s CPI values,[40] and he had other income from his tutoring fees. But Rutherford hesitated. His Exhibition scholarship would not be extended again, and the Coutts-Trotter money alone was not enough.

This put Rutherford at a crossroads, even as his research into uranium and thorium emissions gained pace. He began looking at a Fellowship with Trinity College, which would give him several years' income and keep him working as a full-time researcher at the Cavendish. But still he hesitated. Perhaps he did not want Mary to go through the same crazy rushes of 'feast or famine' that had coloured his youth, times when even feast years were compromised in favour of the Rutherford family's industrial enterprises. What he really needed, as he told his mother, was 'coin, and an unlimited prospect of work.'[41] In short, an academic job. There was speculation that a physics department might be set up in Victoria College in Wellington, for which Rutherford would likely have been a front-runner. But then in April 1898 he heard of a job in Canada. The MacDonald Chair of Physics at McGill University in Montreal, capital of Quebec, had become vacant. The salary was relatively paltry at £500, less than Rutherford expected to get in Wellington and not a lot more than he was already making. But it offered steady work, had excellent repute, and the physics laboratory had just opened in new premises.

Rutherford's mood turned to despair when he thought of the competition. His school-age struggle to get scholarships and his dismal track record of job applications gave him little reason to be confident. Then there was the issue of teaching, something he knew he wasn't good at. He told Mary he thought his chances were slim. 'I will not go unless J.J. advises me to.'[42] But the prospect of steady income and marriage pulled. So did the idea of a secure long-term job in a well-equipped laboratory. McGill was away from the epicentre of Western physics, but he finally decided he would apply, telling his fiancée the salary would be enough 'to start on'. That was no guarantee. 'You mustn't put any faith in my chances.'[43] He was torn between priorities: he knew what he wanted: financial security, life with Mary, and research opportunities that would make his name. But he was not sure how to get there. McGill offered one path. But so too did the Cavendish.

Then, in early June, Rutherford discovered he couldn't get a Trinity Fellowship. He had not been in the college long enough. That threw focus back on McGill. He scrambled to get an application together. Then, as now, the size of his publications list counted when it came to academic credibility. Rutherford had seven papers to his name, one of them co-authored with Thomson and one — on uranium rays — pending. He threw them all into his application.[44] Thomson wrote a glowing letter. 'I should consider any institution fortunate that secured the services of Mr Rutherford as a Professor of Physics.'[45] Other endorsements came from Robert Ball, then the Lowndean Professor of Astronomy and Geometry; R.T. Glazebrook and E.H. Griffiths among them. Amidst the flourishes and endorsements two words from Griffiths stood out: 'scientific keenness.'[46] There was no question about Rutherford's impact at the Cavendish and on the British scientific community, but Rutherford still did not think it was enough. Three weeks later he wrote to Mary to tell her his chances were 'up the spout.' He was getting his testimonials printed, but 'it appears to me it will be money wasted.'[47] His gloom at his prospects did not deter plans for their wedding if he got the job: the university did not operate between April and September, meaning he could return to New Zealand. He was fizzing about working in McGill's Physical Laboratory, one of the best in the world. But his self-doubt nagged.

In the event Rutherford was one of two candidates interviewed by William Peterson, principal of McGill University, and Professor John Cox, then heading the Faculty of Medicine. Both crossed the Atlantic to check out their candidates. And so, in August 1898, Rutherford was appointed professor of physics at McGill University. He was about a month short of his twenty-seventh birthday. 'Rejoyce with me,' he wrote to Mary, 'for matrimony is looming in the distance.'[48] His colleagues were happy for him, though — as Ball put it — 'sorry to lose you here.'[49]

Ernest Rutherford, c. 1910–15.

Library of Congress LC-B2-6101-9, George Grantham Bain Collection

Michael Faraday (1791–1867), seen here in a mid-nineteenth century daguerreotype. One of Britain's greatest experimental scientists, Faraday established the basis of electrodynamics, kick-starting the physics revolution of the mid- to late nineteenth century.

Library of Congress LC-USZ62-110174, Mathew Brady

Rutherford family at Havelock (left to right): Alice Rutherford, Mary Thompson (cousin), Arthur Rutherford (in front), Ernest Rutherford (behind), Eve Rutherford (in front, wearing white), James Rutherford (seated in chair), Nell Rutherford (standing behind), Ethel Rutherford (in front, wearing white), Flo Rutherford (seated in chair), George Rutherford (immediately behind), Herbert Rutherford (at rear), Martha Rutherford (standing side on), Charles Rutherford and Jim Rutherford. Photograph taken before 1886 by William Collie.

PAColl-0091-2-001, Alexander Turnbull Library, Wellington

Rutherford's Christchurch, early 1890s, taken from the Christ Church Cathedral.

Christchurch City Libraries CCL-KPCD14-0012

Rutherford's laboratory at Canterbury University. This converted cloakroom was selected because its concrete floor was less susceptible to vibrations and did not disturb his delicate galvanometers. The space became known as 'Rutherford's den' and was later preserved.

AAOQ W3872 Box 16 40/181 2, photographer unknown, Archives New Zealand

J.J. Thomson.

Library of Congress LC-B2-5882-12, Prints & Photographs Division

Participants at the first Solvay Conference, held in Brussels at the end of October 1911. Standing (left to right): Max Planck, Heinrich Rubens, Arnold Sommerfeld, Frederick Lindemann, Louis de Broglie, Martin Knudsen, Friedrich Hasenöhrl, Solvay's assistants Georges Hostelet and Édouard Herzen, James Jeans, Ernest Rutherford, Kamerlingh Onnes, Albert Einstein and Paul Langevin. Seated (left to right): Walther Nernst, Léon Brillouin, Ernest Solvay, Hendrik Lorentz, Emil Warburg, Jean Baptiste Perrin, Wilhelm Wien, Marie Curie and Henri Poincaré.

PAColl-6238-32, photograph by Benjamin Couprie, Alexander Turnbull Library, Wellington

Rutherford at McGill University, c. 1908.

Library of Congress, Bain News Service LC-B2- 707-6, George Grantham Bain Collection

Rutherford and Hans Geiger in the Manchester University physics laboratory, c. 1912–13.

PAColl-0091-1-011, Alexander Turnbull Library, Wellington

6 | Tom Tiddler's ground

I have to keep going, as there are always people on my track ... The best sprinters in this road of investigation are Becquerel and the Curies in Paris.

— Ernest Rutherford to his mother, 5 January 1902[1]

RUTHERFORD'S MOVE TO CANADA bought new challenges: the demands of a new job, the unknowns of a new country and — inevitably — the unknowns of the radioactive elements he was exploring. What followed was extraordinary, not least because, as his student and colleague Samuel Devons (1914–2006) later observed, Rutherford began work on radioactivity in late 1898 as a brand-new professor in a field where the superstars were Becquerel and the Curies. They had far greater experience than Rutherford and ready access to radioactive materials, which at the time were expensive and primarily produced in Europe.[2] The knowledge that others were already well ahead of him was not lost on Rutherford. He often called such territory 'Tom Tiddler's ground',[3] a reference to a children's game in which someone standing on a heap of stones tries to repel or catch others as they rush in. Rutherford's field enjoyed conspicuous collegiality, perhaps more so than others, but there was status to be gained from being first to make major discoveries in unknown territory. He viewed the Curies as colleagues, but also rivals.[4]

That status was important to Rutherford when he arrived at McGill, as vital in his mind as his curiosity. He had two ambitions. One — which he confessed to Mary — was to become a Fellow of the Royal Society.[5] This was not easy: candidates had to be nominated on merit by an existing Fellow before they were eligible to apply. Even then there was no guarantee: the Society had to judge the nominee's work to be of suitable calibre, then elect to admit them. The bar was high. Anybody who became a Fellow was, by nature, following in the footsteps of the historic greats — Newton, Halley, Banks and others. By this time there were Royal Societies in other places, including New Zealand and Canada, but to be elected a Fellow with the original, *London* Society was the greatest prize, the highest recognition of standing. Rutherford was keen to get it, and early in 1901 wrote to Thomson asking to be nominated. Thomson was favourable but cautious: he had just put somebody else forward and thought it impolitic to 'run another candidate' associated with the Cavendish just then.[6]

Rutherford's push to add 'FRS' to the letters after his name joined his ambition to be awarded a Nobel Prize. This was a high-status annual award first offered in 1901. The prizes in physics and chemistry were administered by the Royal Swedish Academy of Sciences. The process was again one of nomination and then assessment, and here the bar was even higher.

All of it added up to one thing: Rutherford visualised himself in a sprint. Behind it all, it seems, was a desire to be validated — and in terms that nobody could dismiss. For all the booming confidence with which he engaged people, the speed of his anger when frustrated or when he saw slackness in colleagues, he was also at heart the quiet schoolboy from Havelock, the middle child of a large family who likely had to compete with the might-have-beens his mother imagined for his drowned brothers.

Rutherford's final weeks at the Cavendish were busy. He had his work on uranium to wrap up, packing to do and arrangements to make, all in the blazing heat of an unusually long summer.[7] One of the key issues was securing a supply of radioactive material for his experiments in Canada. Germany was emerging as a key player in this market because its industrialisation had generally leaned towards chemical industries, and some manufacturers already had uranium processing facilities thanks to a long-standing use in coloured glassware.[8] Rutherford ordered uranium and thorium oxides from the Görlitz-based concern of Theodor Schuchardt.[9]

Rutherford left for his new job on 8 September 1898, travelling saloon class aboard the steamer *Yorkshire* with Cox and his family, sharing a cabin with Ernest MacBride, McGill's professor of zoology.[10] Rutherford's new job gave him the title 'Professor', but he had yet to take it up, so he was 'Mr Rutherford' on the passenger register.[11] The hurricane season was boisterous that year, and he reached Quebec City on the 18th after a rough passage in vile weather that delayed them a day.[12] The steamer then took to the St Lawrence, a 'magnificent river very wide.'[13] Two days later he was in a freezing wet Montreal.[14] It was a shock. Montreal, too, had been enjoying a heat wave during the first weeks of September, but as Rutherford arrived temperatures plunged. An unprecedented hailstorm swept across the city around the time he arrived, downing a trolley-bus line and killing one pedestrian.[15]

Rutherford was brimming with enthusiasm, his mind filled with questions about the work he had brought with him from Cambridge. He wanted to find out more about alpha and beta rays, and quickly. Becquerel had already discovered that beta rays carried a negative charge and could be diverted by a magnetic field, suggesting they were likely to be Thomson's electrons. But if beta rays were electrons, what were alpha rays, which were far less energetic? And where was the energy to drive them coming from? Rutherford expected such questions would yield to hard work. They also gave him direction when meeting one of the main briefs of the job, expanding the role of the physics department. The radioactive ores he had ordered via the Cavendish had still not shown up, and he wrote back to his former colleagues pressing them to forward the material as soon as it arrived. He obtained a further supply of uranium and thorium oxides the same year from the New York-based laboratory and chemical supplier Eimer & Amend.[16]

The shift to Canada exposed Rutherford to a third culture in as many years and challenged him on multiple levels. The country was larger, more populous, richer and socially more mature than New Zealand, with a legal standing that New Zealand lacked: it was a self-governing Dominion. Like the United States, Canada spread west during the nineteenth century, settlements broadly contemporary with New Zealand, but the eastern regions had deeper colonial history. Urbanisation was pressing ahead by the 1890s. By this time one in three Canadians lived in cities. Industrialisation was spreading in the eastern provinces.

Rutherford's destination, Montreal, stood beside the St Lawrence River and was the capital of Quebec, a region explored and settled during the seventeenth century by the French. It became British in 1760 during the Seven Years' War, but French remained the official language and French culture suffused the region. This history was reflected in clearly defined social enclaves: McGill, for example, was an English-speaking university. The nearby Université Laval, by contrast, was French. Montreal itself was thoroughly modern. Gas lighting reached the city in 1838, the telegraph in 1847. The city was connected by rail to Toronto in 1856, established a fire department and waterworks in 1862, then set up a metropolitan railway that carried just under 1.5 million passengers in 1864. The rate of technological uptake continued to the end of the century and beyond, largely contemporary with the New Zealand experience, but given impetus by proportionately larger population. Electric lighting arrived in 1882, electric trams in 1892, and Canada's first movies were shown in the city in 1896, although it was another decade before the city's first cinema opened.

McGill University was founded in 1821 after Montreal-born merchant James McGill left £10,000 and property to set up a college. Classes began in 1829 and the university flourished over the next decades. Its influence extended into sports: one student, James Creighton, devised rules for a new ball game in 1875, giving the world hockey. Women were allowed to study from 1884. The institution became McGill University the following year. By the 1890s the university was well established in its Montreal campus, set out between Mount Royal and Montreal's downtown district. Rutherford arrived just as the physics department began operating from a purpose-built facility, the Macdonald Physics Building.

All this combined to create a very different environment from Christchurch, different again from Cambridge. His setting — working in a major laboratory with a research team to direct, framed by a substantially different culture from anything he had been exposed to before — offered a new framework for thinking and working. He had a relatively free hand when it came to direction. Cox — as Rutherford's boss — felt that all the main physics discoveries had been made: further work was merely gap-filling.[17] On the other hand, 'rays' and associated radioactivity hinted at a wider vista, and he knew this was Rutherford's strength. He organised Rutherford's workload to minimise teaching — a major blessing from Rutherford's perspective. Even so, he was still expected to have students. He began lecturing on magnetism at the beginning of October,[18] but the results were disastrous. Before long his students were devising rude songs about him.[19] This was not Rutherford's only challenge. He was now in charge of a major laboratory and his sole experience was what he had learned under Thomson. Hugh Callendar, his predecessor, had a huge reputation, and Rutherford finally confessed to Mary that he had become 'rather tired of people telling me what a great man Callendar was.'[20]

In short, Rutherford had to grow into the job, a journey entwined with the mysteries of human nature and given complexity by his youth. He did not look like the stereotypical craggy-browed, white-haired professor and was sometimes not taken for one. His sense of mischief did not help. William Macdonald, the departmental benefactor, had made his money from tobacco but disliked smoking. Rutherford found out, with the result that whenever he knew Macdonald was about to visit he would rush about outside the laboratory building with his pipe, blowing smoke into the open windows.[21]

Rutherford's first major project drew in the American physicist Robert Owens, then the Chair of Electrical Engineering at the university. Owens had research funds from Columbia University and was wondering what to look into. Rutherford had briefly tested thorium at Cambridge and noticed the results were erratic. He couldn't explain this and suggested Owens repeat the trial. Owens did so using Rutherford's equipment design and methodology, including his focus on enumerating any discoveries.[22] Along the way Owens discovered, by accident, that the variable results in the thorium experiments were due to drafts: something was being blown about inside the apparatus. Rutherford could have taken part-credit for the paper that followed. But he did not. This became a pattern: Rutherford would show colleagues or students how to achieve results, coach them, keep an eye on what was happening, and offer guidance. Otherwise they were left to it — and got due credit. As the British physicist Harold Robinson (1889–1955) remarked later, Rutherford was 'extraordinarily generous as a collaborator in, and director of research'. If he appeared as co-author it generally meant he had conducted a 'large personal share' of the work, and 'many papers on which his name did not appear were full of ideas he had freely given.'[23]

This mirrored the technique Rutherford learned from Thomson — but with a twist. Rutherford was ambitious. He kept asking questions, pushing experiments in directions that might — or might not — produce outcomes and then focusing his own attention on what emerged. Any mystery had to be answered, irrespective of how irrelevant it seemed. That became clear during Owens' work on thorium. Rutherford routinely wandered about the laboratory with a pipe in his mouth and found that if he blew smoke into Owens' apparatus it stopped working. As his biographer John Campbell has pointed out, Rutherford wanted to know *why* this happened.[24] One of the outcomes was his discovery that ions were blocked by particulates in smoke. This was later adopted as the principle behind ionising smoke detectors.

Rutherford's first major question was deceptively simple: where did the energy come from? The starting assumption was that internal movement within the atom provoked movement in the aether.[25] In short, radiation was atomic. This was a logical outcome of Thomson's discovery of the electron, and differed from the way the French looked at it. Becquerel still thought radiation was a kind of phosphorescence, while the Curies suggested that the invisible and all-pervasive aether interacted with the heavier elements, provoking them to radiate

RUTHERFORD AND IONISING SMOKE DETECTORS

ERNEST RUTHERFORD'S ACCIDENTAL discovery that smoke prevented ions flowing through his electrical detector led to one of the ubiquitous products of the late twentieth century: ionising smoke detectors. The first was devised in 1939 by Swiss physicist Ernst Meili on the principles discovered by Rutherford. Commercially-made ionising smoke detectors were common by the later twentieth century and among the few consumer products to contain radioactive material, typically americium 241, an alpha-emitter first synthesised in 1944. The alpha particles ionised the air, allowing current to flow between two electrified plates. If smoke interfered with the ions and broke the circuit an alarm sounded. This meant that many households ended up owning what was, in effect, a commercialised version of Rutherford's laboratory apparatus.

By the early twenty-first century these devices were largely supplanted by optical smoke detectors. These exploited the photoelectric effect — something discovered by Hertz in 1888, then explained by Einstein in 1905 — and did not require any radioactive component.

particles. Rutherford's equipment also differed from that used by the French. He disliked their photographic detectors because they did not quantify the results and were insensitive to some of the rays. His electroscope, by contrast, gave prompt results and could differentiate between the emissions of various materials.[26] He had found a way of improving the sensitivity of a standard electroscope by a factor of 100, which was good — though he realised it was still not enough.[27]

Rutherford spent the rest of 1899 and the early months of 1900 working on thorium oxides, focusing on what he called 'emanations' wafting about inside the equipment. He knew Marie Curie had found polonium and radium in pitchblende. Now it seemed that thorium compounds also carried hidden materials. He devised a new experimental apparatus to blow clean air across the oxide, again delivering ions into the electroscope. He discovered that the charge fell by half over a period of eleven and a half hours, dropping in a clean geometric curve. This was very different from uranium.[28] He also discovered that thorium compounds had different properties from thorium itself.[29]

These were extraordinary findings, which Rutherford published in a series of papers, during which he essentially defined what became a standard term in the field, 'half-life', referring to the period of time during which the radioactivity of any active material dropped by half.[30] The chemist Sir Henry Tizard, later, felt that none of Rutherford's papers were more indicative of his genius, given 'his age, the baffling nature of the subject, and the equipment at his disposal.'[31] There was only one downside: Rutherford was 'scooped' by the Curies. Late in 1899 he was in the process of submitting a paper for the *Philosophical Magazine* — Britain's go-to scientific publisher — on the way thorium could induce radioactivity in other substances. This was a brand-new phenomenon and Rutherford hoped to get the kudos for discovering it. But then he fielded a copy of the French scientific magazine *Comptes Rendus de l'Académie des Sciences*.[32] Pierre and Marie Curie had beaten him, finding the same phenomenon in their work with radium and uranium.[33] Rutherford's paper, appearing four months after the Curies' analysis, became the first to report the behaviour of thorium — but not the phenomenon of induced radiation itself.

Science was not Rutherford's only priority in Canada. Mary was on his mind from the moment he set foot in the country. He now suggested to her that they wait 18 months, enough to find his feet. Eventually they were able to sort out plans — all of which Rutherford fitted around work that, as he told her, kept him 'going steadily ... five nights out of seven, till 11 or 12 o'clock.' He was trying to mop up as much as he could 'so that I can take a holiday with a clear conscience.'[34] Rutherford's colleagues had their own ideas when he told them, visualising New Zealand as a relaxed South Pacific paradise of palm trees and the 'dusky maidens' of period colonial fantasy. He found it hilarious and warned Mary that she would be expected to speak in her 'native' language.[35] He also told her he did not want a public ceremony.[36] By December he had firm plans and expected to spend enough to 'travel comfortably on our honeymoon', leaving 'about 1000 dollars in hand when we arrive for furnishing etc.'[37]

Rutherford left Canada in April 1900, a little ahead of McGill's summer close-down.[38] He reached Auckland aboard the *Moana* on 8 May,[39] took a steamer to Taranaki to see his parents, then made his way to Christchurch. He did not tell reporters he was coming, but newspapers were the social media of the day, filled with international news 'by cable', local stories, advertising and social columns devoted to the doings of their local communities. The *Opunake Times* reported that he was at Pungarehu as early as 11 May.[40] The *Star* blew the whistle on his presence in Christchurch,[41] and a *Press* reporter caught up with him on 22 May. Rutherford agreed to discuss prospects for New Zealand students wanting to research in Britain, but as the reporter delicately put it, the 'conversation drifted to other matters, such as the more recent advances in electrical science.'[42]

The moment is telling. Rutherford did not have time to explain the whole field and said so — all he did was give the reporter a couple of points, clearly the ones uppermost in his mind just then. The first was the huge energies involved in radioactivity. The particles emerging from his apparatus, he explained, were moving at about a third the speed of light. One of them, he told the reporter, would be heated to a million degrees if it could yield the whole of its energy. On top of that was the mystery of radioactive materials. Their energy packed a punch and clearly had to come from something inside the atom. The reporter made little of it: a passing remark in a newspaper that would itself be reduced to fire lighter or cut up for toilet paper a few days after being read. But from the historical perspective this was a remarkable statement. Rutherford is usually credited as the first person to openly talk about the colossal energies inside atoms, typically dated to a 1903 paper he wrote with the chemist Frederick Soddy.[43] The idea became a theme in much of Rutherford's subsequent work of the period, notably a 1904 paper discussing the energies involved in radium.[44]

However, the interview in Christchurch makes clear the question of energy was in Rutherford's mind as early as 1900, and must have been for some time prior. This was unsurprising: Rutherford knew the energy he was putting into his laboratory equipment and could calculate what was being carried by the particles coming out. He was also aware of Thomson's conclusion that an electric charge increased the inertia of a particle.[45] He was not the only one thinking along these lines. The extraordinary part was the second half of his remark. In 1900 Rutherford had yet to experimentally show the driving cause for radioactivity. Yet he was still able to tell a reporter that he thought the energy came from inside the atom. In short, Rutherford clearly suspected there was more going on inside atoms than anybody knew.

Rutherford's next remark was equally telling. At the time atoms were regarded as immutable, but he told the reporter that he and his colleagues knew elements were being transmuted during ionisation experiments because they were losing electrons. The quantities were tiny, and to Rutherford that opened up wider prospects: 'It is not scientifically impossible that the old idea of the ancients regarding the transmutation of metals may some day be a realised fact.'[46] It was an astonishing remark. Transmutation was anathema to period chemists, discredited magic — and yet here was Rutherford, openly declaring it was possible. Just a few months later he admonished Soddy for daring to mention 'transmutation' aloud in the laboratory.[47] But his admission to the *Press* reporter makes clear this was where his thoughts were going. Even though Rutherford bowdlerised the term as 'transformation', atomic transmutation was indeed key to his explanation of radioactivity.

Science was on Rutherford's mind in other ways that southern autumn of 1900. He submitted a collection of papers to the University of New Zealand by way of doctoral thesis. He was awarded it the following year, adding DSc to the string of letters accumulating after his name.[48] At the time, a bachelor-level degree was not common, a Master's degree unusual and doctorates were rare.

On 28 June, Ernest Rutherford married Mary Newton in a private ceremony at St Paul's church in Papanui, just outside Christchurch. Reverend H.T. Purchase presided.[49] Rutherford managed to keep his wedding mostly out of the media, though his nuptials were reported by the *Taranaki Herald*.[50] George Rutherford was best man. It was a small affair with just family present: as Mary put it when inviting Rutherford's parents, 'very short and comparatively uninteresting.'[51] The couple spent the next fortnight farewelling family in Christchurch and Taranaki, then departed for a honeymoon in North America. It was the start of Mary's new life as the wife of a professor. For Rutherford the long-held dream of marrying Mary and setting up life with her had come to pass. He found a place to rent near the university, a terraced house with small garden.[52]

Married life was likely a learning curve. There was a gulf between 'dating' and the practicalities of living together. Rutherford's time as a boarder in Mary's family home likely helped. They settled in. Mary gave birth to a daughter at the end of March 1901. Rutherford's colleagues thought the child should be called Ione, after her father's 'respect for ions in gases.'[53] In fact the doting parents named their child Eileen Mary. Rutherford hired a maid. All this was a heavy drain on a salary that had never been fat by the standards of the day. Rutherford was shortly voted an increase by the Board of Governors.[54]

Married life and a young family became backdrop for work that unveiled the explanation for radioactivity. As Rutherford had revealed in that interview in Christchurch, he already had an idea quietly in mind, heretical and radical though it was. The only downside of Rutherford's new job was that it was well away from the mainstream. Despite their rivalries for the limelight, the physics community in Britain and Europe was small, tight, and routinely swapped information and material with each other. Rutherford felt his isolation in Canada, complaining to Thomson that he felt 'rather out of things' and 'greatly miss the opportunities of meeting men interested in physics.'[55] But he was able to find a small team to support his work. He also struck up a brisk correspondence with an American geologist, Bertram Boltwood (1870–1927), a Yale graduate who became, de facto, part of the Rutherford in-crowd during the period.

One of Rutherford's key associates at McGill was his former student, 25-year-old Harriet Brooks. Women were rare in the sciences, and yet here — at the cutting edge of research into radioactivity — were two: Brooks and Curie, both of them talented physicists. Brooks hailed from Exeter, Ontario and enrolled at McGill in 1894. She was declined entry to the medical school because she was a woman, and instead began studying for a Bachelor of Arts. Even that was not easy: women's classes were segregated. Brooks graduated in 1898, studying postgraduate under Rutherford.[56] By 1901 she was in the process of completing her Master's degree in electromagnetism. Rutherford took her into the team.[57]

Rutherford also needed chemists. He initially asked his friend and colleague James Walker, an organic chemist, but Walker declined.[58] Instead he struck up an association with Frederick Soddy (1877–1956), an Oxford chemistry graduate who arrived at McGill by accident in 1900. Soddy explained later that he was looking at a vacant chemistry professorship in Toronto. He didn't get it, but instead found a position at McGill, where the Macdonald Laboratory was looking for a chemistry demonstrator.[59] Rutherford was away in New Zealand when Soddy arrived, and when Rutherford returned, they clashed at a debate set up to discuss Thomson's corpuscle theory. Thomson's discovery was still controversial and Soddy was firmly of the Daltonian school, declaring that atoms were not divisible. Chemistry, as understood at the time, could only work if they were solid. Rutherford knew otherwise, remarking to Thomson that: 'we hope to demolish the chemists.'[60]

Soddy had been given little to do and his cynicism soon gave way to curiosity: he found himself spending time in Rutherford's laboratory.[61] It was informal at first, but the head of department, Bernard Harrington, did not seem to object.[62] Finally Soddy became a convert and threw himself into Rutherford's ambit. It was a decisive shift. As Soddy remarked later: 'For more than two years life, scientific life, became hectic to a degree rare in the lifetime of an individual, rare perhaps in the lifetime of an institution.'[63]

The drive came from Rutherford. He worked all hours and at times even showed a competitive streak with his own team. He knew his British and French colleagues were pursuing the same goal and that the Curies, in particular, were well ahead. The secret of radioactivity was going to yield to somebody, and soon. Whoever got there first would gain high standing, and Rutherford had both a Fellowship of the Royal Society and a Nobel Prize at the back of his mind. As one analyst has shown, that gained force from the fact that the Curies had already beaten him to the punch in 1899–1900 over radioactive 'excitation.'[64] We must also add his experience as a graduate student in Christchurch, where a British team published on his subject before he could even begin work.

That early experience highlighted what to Rutherford's mind was the real challenge of the new science race. Whoever got into print first got the credit. Rutherford had already been 'scooped' by the Curies in 1899–1900, and there was no prize for second place. In a practical sense this stood against the reality of a field where multiple researchers were discovering incremental aspects of the bigger picture, frequently with overlaps, often collaboratively. But 'first-in, first-acclaimed' was how the status ladder worked, closely keying into popular thinking that threw a spotlight on any 'first', irrespective of nuance. Rutherford was acutely aware of the issue. It was on his mind to the point where — as one example — he felt obliged to remind readers of one paper, that he had been first to discover that radioactive materials 'emanated' an unknown gas, ahead of the Curies.[65] Publishing was also a conversation. As Pierre Radvanyi has pointed out, Rutherford effectively engaged in an academic discussion with Pierre Curie in particular over the nature of radioactivity, played out in journals.[66]

Rutherford had two issues when publishing. The first was time. He was known for carefully expressing himself in papers,[67] a clarity that stood at odds with the vagueness of his conversations. Those who got to know him eventually realised he had a clear vision in mind. Rutherford's struggle to translate his ideas into the linear thread demanded of written words is clear from handwritten drafts of the papers he wrote at McGill, filled with strikethroughs and recastings.[68] He gave these to Mary to type them for delivery. Yet even clean typed drafts of his papers from the time carry pen-and-ink amendments, in part because typewriters didn't have specialist mathematical symbols, but also because he was still adding data and adjusting nuance.[69] For Rutherford, then, translating the dimensional concept in his mind to words was not quick even with the input of co-authors. Even then, as a brief controversy that erupted in 1906 reveals, he felt his words were sometimes misunderstood.[70]

Having a finished manuscript in hand was only half the story. Rutherford's go-to publisher to this point was *Philosophical Magazine*, established in 1798, and a significant outlet for much of Britain's nineteenth and twentieth century science. Its editorial board of Rutherford's day was led by William Thomson, Lord Kelvin, the grand master of Western physics who did not let his cynicism about some of the new science get in the way of giving newcomers a voice. The magazine published monthly, but the system was wrapped in checks and balances, effectively an early form of peer-review. Any findings had to be first 'communicated' — meaning either presented by lecture to a recognised organisation, or sent to somebody involved with one of the major laboratories. Only then could a paper be submitted.

Rutherford's location in Montreal added further delays. His manuscripts had to be sent to a colleague in Britain — often J.J. Thomson or Crookes — then independently checked, returned to Rutherford, and finally submitted. Then there was the publishing process itself, including author proof-checks. There were delays of three or four weeks each time while packages were shuttled across the Atlantic and back — weeks of delay that authors in Britain did not experience.[71] As just one example, Rutherford's first paper on thorium emanation — the first to publish the decay-curve chart for radioactive materials — was finished in mid-September 1899. Thomson turned it around promptly, for which Rutherford thanked him,[72] but even so the paper did not appear in print for five months.[73] Indeed, there was no guarantee of prompt release even when the paper had gone through all the steps. One of Rutherford's joint papers with Brooks, for example, was based on research conducted in 1900, but did not appear until 1902.[74]

Meanwhile plenty of other people were working on the same thing: Becquerel and the Curies in Paris, William Crookes and others in England, Friedrich Dorn and others in Germany. When it came to publishing there was no way Rutherford could compete with the Curies, who had a finely honed sense of self-promotion and whose go-to magazine *Comptes Rendus* was geared to get information quickly into the public arena. The physicist Lawrence Badash (1934–2010) has noted the difference between the Curies' material and the more carefully organised outputs of Rutherford.[75] At the time the French magazine was, in effect, the early twentieth century version of online pre-print outlets such as Cornell University's online arXiv repository.[76]

In 2021, the US physicist and science historian Melinda Baldwin showed that Rutherford found lateral answers to the timing problem, pushing material in *Nature*, a weekly science magazine then pitched for the popular market. For Rutherford the attraction seems to have been speed: as one example, a letter he wrote at the end of May 1901 on 'emanations' was in print just 14 days later.[77] Baldwin has argued that Rutherford's material helped transform *Nature* into a credible scientific journal.[78]

In fact it seems Rutherford used *Nature* as a pre-publication outlet, getting a summary of his next paper out so it could not be scooped, then issuing that paper with its mathematics and data in a more mainstream journal of the day. He made a point of building good relationships with publishers, making friends of the key figures. When Rutherford sent a paper of Arthur Eve's to the *American Journal of Science* he knew it would be well received by the editor, Edward Dana. 'The paper is, of course, most welcome and will find a place in our coming December number,' Dana responded before launching into a chatty note that discussed everything from the German physicist Walther Nernst's thickly accented English to Dana's own stomach problems.[79]

Timing was only one aspect of the publishing problem from Rutherford's perspective. He also managed where, when and how he published, and in ways intended to make sure his rivals knew it directly. This became clear when he restarted work in September 1900. Rutherford's priority in the new university year was the 'emanation' he and Owens had first identified from thorium oxide, which had the 'remarkable property' of 'causing all bodies in contact with it, to become themselves radio-active.'[80] Rutherford had been able to concentrate the effect electrically, forcing the emanation to gather around the negative electrode in his apparatus. This also meant he could measure the length of time it took for the charge to fall away — which in this case was about 12 hours.

Where next was driven by developments in Europe, where the Curies and their colleagues had turned to radium, a violently radioactive alkaline earth metal that — thanks to that activity — had potential to provide a lot of data. It was typically produced as a salt, radium bromide. However, the concentrations of radium in that compound varied between makers. Rutherford had a small supply from the German firm Estler & Geitel. He knew it was not particularly concentrated, but it was all he had to hand. He could not detect any 'emanations' from it, but then learned that the German physicist Friedrich Dorn had obtained results from radium sourced elsewhere.[81] For Rutherford it was another blow: the Curies had beaten him in the race to announce *induced* radioactivity, and now he had come second when it came to emanations from radium, let down by the poor quality of the material he had in Canada.

Rutherford bought more radium from Dorn's source, the Chemische Fabrik of E. de Haën. This plant had been founded near Hanover by Carl de Haën in the 1860s, producing uranium salts at industrial scale for use in coloured glassware. Their radium bromide was high quality, and once Rutherford had some the pressure was on to do something Dorn hadn't. Earlier on, Rutherford had found that by heating thorium to a red glow he could increase its emanation, but once it was white hot the emanations stopped. Now Rutherford decided to heat some of the de Haën radium. The results were decisive. He had not warmed the sample far before it went mad, spewing ten thousand times the emanation of cold radium.[82]

To Rutherford that posed more questions than it answered, but he was determined to get the word out before Dorn thought of carrying out the same experiment. This took the form of a paper in German submitted directly to Dorn's go-to magazine, the periodical *Physikalische Zeitschrift* (*Physics Journal*). Rutherford's piece was published as the lead article in the April 1901 issue, and he gave his own discovery context by prefacing it with Dorn's work on emanations from polonium, thorium and radium. Rutherford had learned German during his 1895 voyage to Britain, when he did not know whether he could get into Cambridge and saw prospects in Berlin as a backstop alternative.[83] Even so, he had to go to considerable effort to translate the paper, more than had he simply issued the finding to *Nature* or the *Philosophical Magazine*. It was, in short, a carefully targeted shot across Dorn's bows and a statement of position, one he knew would also be read by Dorn's colleagues.[84]

Rutherford's next focus was the 'emanations'. He devised the generic term deliberately because he had not been able to gain any clue as to what they actually were — whether they were vapour emerging from the radioactive material, a specific gas, or merely particles.[85] There was not enough to examine chemically. However, the quantities he could get by heating samples of his radioactive materials were enough to yield to his trusty electromagnetic detectors. By contrast with thorium, radium emanations remained radioactive long enough for him to measure the rate at which it diffused through air. This meant that he could determine whether it was a gas or not.[86] That same experiment would also yield the molecular weight, using a diffusion formula developed in 1871 by the Austrian chemist and physicist Johann Loschmidt (1821–95). Rutherford also knew that once he had that information, a lot of other data automatically followed. This was yet another instance of the principle that had driven the Royal Society to observe the transit of Venus in 1769: one key piece of data could unlock a plethora of other information. But by contrast with astronomy — where the pieces of the puzzle were known — Rutherford's task involved first discovering enough to deduce what the pieces were.

Rutherford called Brooks to work with him on the problem. This required yet another ingenious piece of equipment, a classic Rutherfordian recipe founded in the techniques he had learned at Canterbury. The heart of the device was a diffusion tube devised by Loschmidt. This comprised a brass cylinder divided by a movable metal slide, to which Rutherford added his electrometers and a system for heating the radium. Experiments began by blowing the emanation into one half of the cylinder. Once an electrometer showed that enough mystery material had been collected — a process that took an hour or so — the cylinder was sealed and the slide opened, letting the emanation diffuse. An electrometer attached to that side let Rutherford and Brooks measure the rate of spread. It was not quite as simple as it sounded: Rutherford knew the emanation would induce radioactivity in his detectors, skewing the results. To get around this he quantified the skew with an additional detector plate he could replace just before taking primary measurements. The difference between old and new gave him the value of skew.

Brooks and Rutherford discovered that the emanation was a radioactive gas and gained a broad indication of the molecular weight, which was somewhere between 40 and 100.[87] This was a major step. The molecular weight carried wide uncertainties but still lay well outside the molecular weight of radium vapour. In short, it was a different element. The new gas remained radioactive far longer than the emanations of thorium. Rutherford now wanted to test how this changed over time and how that affected the results. One of the new questions puzzling the community was why emanations faded at varying rates, sometimes so quickly it was difficult to measure. Rutherford noted that the diffusion experiments he and Brooks were making with radium would not have been possible with thorium for this reason.[88]

WHO DISCOVERED RADON?

THE RADIOACTIVE EMANATION discovered by Dorn and then explored by Rutherford and Brooks eventually turned out to be a new element, an inert gas in the same series as argon. After some to-and-fro discussion within the scientific community it was eventually named radon. There has since been debate among historians as to who first identified it. For a long time the discovery was credited to Dorn, who announced it in 1900 and whose claim became embedded in the scientific literature. However, an investigation in the early 2000s showed that Dorn had been prodded down that line by Rutherford's own earlier work, and had not strictly identified it himself.[89]

According to this study, the main work underpinning the discovery was undertaken by Owens, at Rutherford's urging; and then by Rutherford himself, whose studies of thorium in 1899 produced what — with hindsight — we now know to be the short-lived isotope radon-220.[90] Almost simultaneously the Curies discovered what we now know to be radon-222, a longer-lived isotope emitted by radium.

Rutherford's confirmation that the emanation was a gas did not slow the race. The Curies, with the support of the chemist André-Louis Debierne, also found that radium produced 'emanations', publishing their results on 25 March. Rutherford responded promptly. Instead of waiting to expand their results, he and Brooks assembled a paper reporting just their first findings, which was read on 23 May and submitted to the *Proceedings and Transactions of the Royal Society of Canada*.[91] But that was still going to take time, so Rutherford also wrote a two-page letter to *Nature*. This was a further shot across the bows of his colleagues in Europe. His letter drew much of its wording from the paper in press with the *Proceedings*, but was published well ahead of it on 13 June.[92]

Once again Rutherford gave place to his work, reminding readers of the Curies' discovery that radium produced a gas that remained radioactive for minutes, explaining that he had done the same for thorium, and that he had then shown how he could increase the output of radium by heating. All this was a prelude to summarising his discoveries with Brooks and his declaration that the emanation had to be a radioactive gas, though he had insufficient space to explain how it all worked.[93]

To issue this in advance of the paper — with his hint that he knew what was going on — makes clear enough what Rutherford was up to. The assertion of place was not subtle, and his selection of *Nature*, with its weekly publishing schedule, underscores his urgency. None of this reduced the fact that Rutherford's field was collegial, perhaps more so than other branches of the sciences and humanities. The academy in general had a reputation for senseless rivalries. The classic period example was the way Sir Richard Owen bullied Gideon Mantell over dinosaurs in the mid-nineteenth century, a struggle that did not end with Mantell's untimely death in 1852. Mantell had suffered from scoliosis and Owens acquired his rival's damaged spine, which he had preserved and kept in the Hunterian Museum, ostensibly for scientific purposes.[94]

Western physics at the turn of the twentieth century was never so vicious. For all the debate over findings, and for all the race to find answers first and earn the glory, the physicists working at the cutting edge of the field were a tiny community. They generally knew each other directly or through correspondence, often actively forging friendships at a distance. 'My dear Professor Rutherford, I beg leave to propose that we drop titles in addressing each other,' the Dutch physicist Heike Kamerlingh Onnes (1853–1926) declared in one letter.[95] They gossiped, including when there wasn't anything to say — as when the physicist H. Lester Cooke (1879–1946) told Rutherford that he was writing because there was no news.[96] Academics in the field shared ideas, resources and equipment and encouraged each other.[97] 'I am very much obliged with the trouble you took,' Cavendish student R.K. McClurg told Rutherford after the latter read a draft paper, 'as it was my first attempt at writing up such a paper for publication.'[98] Rutherford pushed the field: his energy lasted among the audience, a point noted by the US physicist and inventor Fay Cluff Brown (1881–1968) after Rutherford visited the University of Illinois in May 1906. Rutherford was particularly supportive of friends such as Bragg: 'I hope you will publish your work in the Phil Mag as soon as you can.'[99]

At other times one physicist would ask another to check their findings. If new gear came along they were quick to tell those who might use it, as in 1903 when Cooke was shown a new electroscope design and promptly told Rutherford.[100] Others were generous with their interests. When the US physicist Clement Child discovered that his work on the velocities of ions overlapped Rutherford's he offered to 'resign that work in your favour' if Rutherford had plans to explore it.[101] Rutherford, for his part, was unstinting with the credit he gave to associates such as Brooks, Soddy and later Hahn and others. They knew it, too. Brooks even wrote to Rutherford asking him 'not to give me any more credit than I deserve ... you are quite too generous in that respect.'[102] Rutherford also kept up regular correspondence with the Curies. They shared the sentiment, and later in 1901 Marie Curie sent Rutherford high-quality radium from her own supply.[103]

Despite the pressure to be 'first', all understood that the end-goal was ultimately a shared one.

Rutherford's problem was that his job in Montreal was on the periphery: he could keep in touch by mail but had few opportunities for direct discussion. The face-to-face meetings he could get were often time-consuming to obtain: in 1902, for instance, he took the train to Washington for a meeting of the Physical Society. It was 'rather a long journey' but — as he told Thomson — 'one way of keeping in touch with what is going on down south.'[104]

The 1901–1902 academic year started in September, and Rutherford got cracking on a new direction. He was joined by Soddy, who primarily worked from the chemistry laboratories. This time Rutherford wanted to define emanations from thorium. It was challenging: the quantities they were working with defied chemical analysis and were difficult to detect even with his electroscopes. Then Rutherford hit upon the idea of cooling the emanation. An early experiment with dry ice failed. Sir William Macdonald provided money to buy equipment to produce liquid air. This meant Rutherford could get down to temperatures of around -194° C (-317.2° F), well below the liquefying temperature of oxygen and just above that of nitrogen. Because of the differing temperatures at which air's two main gases liquefy, the resulting mixture was just over 50 percent oxygen, more than double the usual atmospheric concentration.[105]

Later, Soddy recalled that the flurry of work made one moment blur into another. It was difficult, later, to 'single out the steps' in his memory. He recalled the moment when the emanations were condensed with liquid air. Another key moment came when they discovered a substance Rutherford dubbed 'thorium X.'[106] A few years earlier Becquerel had separated a material he called 'uranium X' from uranium, reducing the radioactivity of the original uranium sample to undetectable levels. Now Rutherford and Soddy had managed to do the same with thorium, theorising that it was a way-point between thorium and its emanation. But there was a difference. The original thorium retained a large fraction of its radioactivity. This was a new question that demanded answering.

Amidst this work Rutherford turned his attention to wireless again, and also continued to explore alpha rays, which he saw as part of the radioactive puzzle. Along the way he managed to electrocute himself on his laboratory equipment. Soddy found him:

> *dancing like a dervish and emitting extraordinary imprecations, most probably in the Maori tongue, having inadvertently taken hold of the little deviation chamber before disconnecting the high voltage battery, and under the influence of a power beyond his own dashing it violently to the ground, so that its beautiful and cunningly made canalization system was strewn all in ruins over the floor.*[107]

Meanwhile a new figure entered the ring: a theorist, James Jeans, then an Isaac Newton student at Cambridge. Jeans suspected that radiation was related to the structure of matter. He analysed this concept mathematically in terms of existing electromagnetic theory, publishing his study in November 1901.[108] Jeans' work underscored the fact that radioactivity and its causes was becoming a major focus of the field, with implications that ran far beyond simply discovering the properties of radioactive materials.

As Christmas that year approached, Rutherford was still pondering the mystery of thorium and thorium X when he was tipped off to a new phenomenon. He had sent a paper to Sir William Crookes for review prior to publication. He also wanted more thorium salts and knew that Crookes obtained his own supplies from the factory owned by the German chemist Oskar Knöfler. Crookes responded with a generous note typed on his personal letterhead, agreeing to pass on the request and offering some of his own stock if Knöfler did not come through. In some hasty final lines jammed at the bottom of the sheet, Crookes added that Becquerel had asked him to cross-check something odd. The French physicist had been able to separate highly radioactive material from uranium nitrate, apparently removing the radioactivity from the uranium. But then — a little later — he found the inactivated uranium had become radioactive again. Crookes was intrigued. 'I am at work on old compounds of my own to see if I can get similar results.'[109]

Rutherford was intrigued, too. But the Christmas break was upon them. He chafed. 'I have to keep going,' he wrote to his mother on 5 January, telling her that there were 'always people on my track' and that he had to 'publish my present work as rapidly as possible in order to keep in the race.' His key rivals, he admitted to her, were Becquerel and the Curies.[110]

When the university re-opened Rutherford found himself front and centre of the new mystery. He and Soddy had left the samples of thorium and thorium X in the laboratory over the Christmas break. When work resumed Rutherford found the thorium was once again fully radioactive. This was the same phenomenon Becquerel had observed in uranium. Something was up, and it had happened over Christmas, meaning Rutherford missed observing the specifics. There was only one answer, and Rutherford threw himself into fresh experiments. For a while they found nothing. Then Rutherford made a breakthrough. It happened on a Saturday. Soddy left Rutherford working on a deposit that did not seem to be reacting as it should and returned after lunch to find Rutherford pacing about outside the physics building. The deposit, it seemed, was producing increasing quantities of radioactivity.[111]

This matched what Becquerel had found. Rutherford and most of his colleagues had long suspected that radioactivity was carried by the emanation, and that removing this from the source material would inactivate it. But now it was clear something else was going on. For Rutherford it was another piece in a jigsaw where he didn't know either the final picture or the number of pieces. However a pattern was building, and — as he told that reporter in Christchurch in 1900 — transmutation lurked in there somewhere. All the evidence he and Soddy had collected so far, including from work of others such as Crookes, Becquerel and the Curies — pointed to elemental change as the central core of what was going on. It was profoundly heretical to chemists, yet obvious to physicists — and the missing part was a mechanism to link it all together.

The work culminated in a two-part paper that Rutherford co-authored with Soddy.[112] The first part appeared in the September 1902 issue of the *Philosophical Magazine* along with another paper Rutherford co-authored with A.G. Grier on alpha particles, a closely related topic. The hypothesis that Rutherford and Soddy came up with was elegant, simple, and entirely radical, embodying the two ideas on Rutherford's mind back in mid-1900, energy and transmutation.[113] Radioactive elements spontaneously disintegrated, releasing particles and energy in the process, and so transmuting themselves into other elements — the 'X' elements that Rutherford and other researchers had been finding. This process played out to specific timings. To measure that process, Rutherford and Soddy used Rutherford's earlier concept of half-life as a yardstick and — in passing, almost as a throwaway point — noted that the radioactive ores being found might require extended geological time to form measurable quantities.[114]

At this stage the idea of radioactivity as a product of atomic disintegration was a hypothesis — a proposal, not yet a theory, demanding a wider range of proofs than Rutherford and Soddy could offer even in a two-parter. Today, largely thanks to work Rutherford undertook to follow up his claims, we know that they got it right. Radioactive elements spontaneously break down through a process that transmutes them variously into different isotopes of the same element, or into different elements altogether, known as a 'decay chain'. It was not obvious, and the process of figuring it out was made all the more complex by the fact that at the time Rutherford did not know the mechanism behind it. Worse, some of the steps in the decay chain were fleeting and as a result were missed or conflated. It is a measure of Rutherford's genius that he was able to figure out the picture from the partial chain he produced through experimentation. He also did it without knowing quite what alpha particles were and without being aware of the structure of the atom or its component particles — data that could have told him more about the process. To go from fragmentary evidence to the correct conclusion without having the full breadth of detail was an astonishing achievement.

The other side of Rutherford's discovery was its audacity. In 1902 the idea that atoms were composed of smaller particles was considered true among the physicists, but chemists begged to differ. To then declare that discredited alchemy was not only real, but a naturally occurring process from which radioactivity drew its power, was bombshell territory and an invitation to ridicule. It was all the more so because with this hypothesis Rutherford and Soddy had also declared that the thermodynamic ideas of the Curies, the phosphorescent concepts of Becquerel, and the other notions raised to explain radioactivity — all well-founded in classical physics — were wrong. Why? Because nature was an alchemist. Hard data or not, this was akin to shouting 'magic!' from the top of the Macdonald Physics Building.

Rutherford was well aware of this challenge. He expected an uphill battle and wanted to fire an opening salvo directly into the chemistry community. Early in 1902 he and Soddy prepared a paper for the Chemical Society, but Rutherford was worried that they might reject it as too radical and leaned on Crookes for support.[115] The irony, of course, was that Rutherford's Nobel Prize — when it came in 1908 — was for this very work, and was awarded in chemistry. In the end Rutherford bowdlerised 'transmutation' in his formal publications, calling it 'transformation'. But he used 'transmutation' in private letters and his public word change did not disguise the concept. Still, the message had to be got through and in early 1902, as he assembled the two-part paper for the *Philosophical Magazine* and got his supporting publications in order, Rutherford decided to also write a book.[116]

The division of the main hypothesis into two parts for the *Philosophical Magazine* was another demonstration of Rutherford's mastery of the system, a one-two punch that laid out the key concept — and the message was clear: watch this space. It did not take long to fill. Rutherford pumped out papers with machine-gun rapidity over the next few months, including in Germany. More followed in 1903. Some of these were catch-ups, reports on his earlier researches that were finally getting column space in the journals. Others were new, including his final co-authored paper with Soddy, 'Radioactive Change'. This explored what we now know as decay chains, along with the energy involved in radioactivity.[117] When Rutherford came under fire — as he did early in 1903 when Becquerel claimed that Rutherford's electrical detection methods produced faulty results — he fired back, in that case by letter to the *Philosophical Magazine*.[118] Rutherford felt all of this was necessary because his ideas were radical for the time and — like all radical ideas — controversial irrespective of the degree of proof. The French, in particular, were tough nuts to crack, though his English friends were convinced he was right. Oliver Lodge reassured him that 'in this country your reputation is safe.'[119]

In this circumstance Rutherford's planned book had every potential to add weight to his case, bringing together his discoveries into a single compelling volume. He had it under way by mid-1903, work that continued apace even as he took a mid-year break in New Zealand. He wrote to Bragg from Pungarehu to say how sorry he was he could not visit Australia this time — he was 'fully at work on my book', had 'left a month later than had hoped' and was 'very much rushed for time' due to having to attend a conference in Vancouver.[120]

Rutherford remained aware of the fact that he was but one of multiple physicists looking into the field, and barraged *Nature* with notes, reasserting how he was extending the discoveries of others, or answering questions his colleagues had left unanswered, as in October when he and his colleague Howard Barnes issued a lengthy letter titled 'The Heating Effect of the Radium Emanation'. This was specifically to answer a question arising from work by Pierre Curie, and again reasserted the main deal: everything observed could be explained by his disintegration hypothesis, by which the internal energies contained within atoms were being released.[121] A full paper followed in the *Philosophical Magazine*.

Rutherford's book on radioactivity followed. He asked both his wife and Harriet Brooks to help check the proofs, and it was published in 1904 by Cambridge University Press as part of their Physical Series. He dedicated it to Thomson.[122] This came on top of duties as a university professor, work that now extended to reviewing papers for the major journals.[123]

In and around this Rutherford filled gaps in his theory and detailed his research. He was interested in new forms of radiation and the energies involved in radioactive emission. The best material for this work was radium, but it was hard to get, particularly in Canada. Rutherford was pipped at the post early on by Dorn because his initial stock was poor quality. Rutherford then obtained a small amount of quality radium bromide — with a higher concentration of the element — from Dorn's supplier. Marie Curie's 1901 gift was useful. But Rutherford needed more and at the time the Curies essentially stood atop the main supply.

Then something extraordinary happened. Soddy left Montreal early in 1903 to join Sir William Ramsay at University College in London, primarily investigating the products of radioactive decay. Some months later, Soddy happened to be walking down Mortimer Street, glanced into the window of equipment supplier Isenthal & Co — and found radium bromide on sale. It turned out that Isenthal had secured a supply made by Friedrich Giesel at the Chinin Fabrik in Brunswick and was selling it at 8 shillings a milligram, about £40 in mid-2020s money. Soddy could hardly believe it: here was one of the most expensive and difficult-to-obtain substances on Earth, openly for sale in a London retail outlet. A few months later Rutherford arrived in Britain, met Soddy, and the pair walked to Isenthal's premises in Mortimer Street where Rutherford — as Soddy put it — was 'absolutely bowled over' by the find. He promptly purchased 30 milligrams, which he initially lent to Soddy but which were subsequently sent back to Canada.[124]

Later, with the help of $500 provided by Sir William Macdonald, Rutherford purchased a further 100 milligrams of radium direct from Giesel. This gave Rutherford and his team enough to conduct useful experiments.

Problems obtaining radioactive material remained an issue for the physics community of the day: supplies across the major laboratories combined totalled just a few grams and the quality of the various supplies varied. The Curies had the best stock, enabling what Boltwood called 'wild scientific orgies' in their laboratory, thanks to '280 MILLIGRAMS OF PURE RADIUM BROMIDE'.[125]

Radium came with a number of problems. It was expensive, rare, dangerous and — critically from the scientific perspective — it produced radioactive gas, which was very difficult to contain. Everybody was on a learning curve. The message was brought home to Rutherford in 1904 when his student Arthur Eve could not get an electroscope to work properly. This was no fault of either Eve or the equipment. On investigation it turned out the laboratory had been contaminated with radon leaking from the experiments.

Rutherford took precautions after that to prevent it escaping, keeping his radium stored in an airtight container, in the maximum vacuum he could achieve with his Toepler pump.[126] But in many respects this was a fortunate accident. Rutherford routinely handled mildly radioactive materials, even carrying pitchblende around in his coat pocket. He also stored ores in his desk. One of the outcomes was that some of his papers, both from his time at McGill and his later Manchester days, remain slightly radioactive.[127] However, radium was orders of magnitude more dangerous than any of the radioactive ores — something obvious to the naked eye from the fact that radium bromide emitted a green glow and it produced burns on naked skin. Rutherford had seen the effects on Pierre Curie's hands during a visit to Paris in 1903. The need to handle the McGill radium supply extremely carefully, though motivated by the need to avoid losing an expensive resource and to prevent it contaminating the laboratory, also reduced the risk to the scientists. Even so there were accidents, including an embarrassing moment when Rutherford lost a small phial he was taking to Ottawa for a lecture. It remained somewhere in the railway carriage he had used and would continue to radiate, as he pointed out, for several thousand years.[128]

Rutherford's book was published in mid-1904, cementing his place in the field. The American physicist Henry Bumstead, then teaching at Yale, read most of it within 24 hours of purchase, writing to Rutherford to offer 'warm congratulations.'[129] It was widely read, even by scientists not otherwise involved in that territory. The astronomer Edward Barnard (1857–1923), then working at the Mt Wilson observatory east of Los Angeles, read it and promptly wrote to Rutherford for advice about fogging problems on some of his photographic plates, which he put down to a makeshift repair that — he suspected — had provoked a radioactive release from the metals involved.[130] Rutherford also found himself in demand for lectures to the point where they interfered with his other work and he began turning them down.[131]

He produced an updated edition the following year to incorporate his further researches and findings, notably his later insights into radium decay products. One of them, which he dubbed 'radium F' (polonium-210) turned out to be the variety Marie Curie had first noticed.[132]

Why did Rutherford find the answer and not the Curies? Rutherford had a well-equipped laboratory and talented team to back him, but so did Marie and Pierre, and they were well ahead in the field with easy access to good supplies of high-quality radioactive materials. What were the differences? In 1979, Marjorie Malley suggested the general issue was differing philosophies: specifically, French positivism and their thermodynamic model, versus Rutherford's materialistic realism and his relentless drive for empirical data.[133] It was a fundamental divergence. Efforts to detail the Curies' equivalent hypothesis, though, have been difficult because they avoided detailed theoretical discussion in their papers. In 2005 a study concluded that they had adopted the ideas of French physicist Georges Sagnac (1869–1928), who believed the radioactivity displayed by the elements was secondary: in other words, this material was re-emitting radiation it had already absorbed.[134]

In the end Rutherford was correct. His explanation of radioactivity defined the basis of the new field by establishing a new principle, spontaneous atomic disintegration, a concept with far-reaching implications on a scale similar to Newton's analysis of gravitation. Pierre Curie finally agreed that Rutherford had got it right.

The other question is Soddy's role. He played a significant part in establishing the detailed data that proved the disintegration concept and contributed to the hypothesis as it evolved. However, Rutherford had a working hypothesis before Soddy appeared on the scene: the idea of atomic transmutation and subatomic energies was uppermost in his mind as early as 1900, well before they began their collaboration, though exactly how matters worked were not known to him at the time. He also seems to have taken the lead in the work and in writing it up: as he later told Bragg, Soddy was 'inclined to be very rash' when writing: 'I have always felt Soddy wants a hard-headed man alongside of him to run a blue pencil through undesirable references in his publications and generally to steady him.'[135] Rutherford also did much of the work — again with other collaborators — that detailed the theory he had hammered out with Soddy.

Delivering this new paradigm to the world would have been enough to guarantee Rutherford's place as one of the leading scientists in Western history. But for this energetic New Zealander that was only the beginning. Knowing how radioactivity occurred did not explain why atoms spontaneously broke apart, a question he was determined to answer. And his findings gave pace to a more general revolution in physics that by 1904–1905 was reaching storm pitch.

7 | A mass of energy

ERNEST RUTHERFORD'S EXPLANATION of radioactivity was a pivotal moment in the history of modern physics. For the first time a physicist had fully overturned classical conceits with an irrefutable array of hard data, showing that the new approach — in which everything was reconsidered in terms of electromagnetic theory — was able to produce valid explanations. And he had done so with irrefutable evidence built from painstaking research. In the process Rutherford laid the essential foundations for both nuclear physics and the broader field of particle physics. This was a big deal, and recognised as such almost immediately by colleagues in England. The icing on the cake was his election to the Royal Society of London.[1] In 1904 the Society added the prestigious Rumford Medal to Rutherford's growing list of honours. Only the grand master of British physics, Lord Kelvin, was hesitant to accept Rutherford's ideas. Rutherford was reluctant to respond but eventually felt he had to engage Kelvin over it, carefully and delicately.[2]

Rutherford's European colleagues were also hesitant: Rutherford's idea overthrew older certainties, and he had much work to do filling in the gaps. One of those was the full radium decay chain, along with the precise identity of the particles spat out by the material at each step — which, in turn, was going to reveal much about the internal makeup of the atom. Rutherford knew Thomson's electron was one of those particles, emitted as beta radiation. But the first three steps of the radium decay chain involved an alpha particle, which he had yet to identify. The fifth step involved alpha, beta and gamma particles together.[3] Finding the specific properties of an alpha particle, then, was high priority. By January 1905, as Rutherford told Boltwood, he was 'convinced that the alpha particle is a helium atom with two charges'.[4] He explored that further, including his speculations in his Silliman Memorial lectures, which were published in 1906 as *Radioactive Transformations*. Here he calculated, purely on a theoretical basis, that a gram of radium should produce about 0.11 cc of helium annually.[5] This was verified two years later by the Scottish physicist James Dewar, who ran an experiment that produced practical results close to Rutherford's prediction, and who made clear when reporting the outcome that this validated Rutherford's work.[6]

All of this added to a growing picture of an extraordinary and complex process behind radioactivity — all pointing to a further and profoundly fundamental question. Why did radioactivity occur? Exploring the underlying cause was Rutherford's main focus by this time. The answers clearly had something to do with the structure of atoms, but in order to approach that question he realised that he needed a good deal more data, including proper proofs as to the nature of alpha particles.

Rutherford was joined by new students during 1905, including Otto Hahn, a German chemist who had been working with Soddy in London. Hahn later recalled the experience of working in Rutherford's laboratory. Watching for scintillations demanded darkness, so the key equipment was in a 'dark cellar full of nooks and corners'. Rutherford led with an enthusiasm that 'infected all the fellow workers' and 'made us into one united family'. Hahn arrived thinking a professor was likely to be 'rather unapproachable' but instead found Rutherford generous, kind and unassuming, a father figure who welcomed his students to his home like family. Hahn enjoyed the evenings at Rutherford's house 'listening to Mrs Rutherford's talented piano playing and the Professor's spirited narrative'.[7] He recalled later how Rutherford's 'enthusiasm and abounding energy infected us all and it was the rule, rather than the exception, to work on at the Institute after supper.'[8]

That year was a whirlwind for Rutherford. In March he delivered a broad summary of his thinking in a series of Silliman Memorial lectures at Yale. Here he caught up with Boltwood, socialised with his colleagues, and as he put it, had a 'thundering good time.'[9] The fact of being selected for the prestigious Silliman series underscored the place Rutherford's ideas had in the field. He was paid $2500 and the deal included publication.[10] Amidst that work he decided to update his book on radioactivity, eventually expanding it by almost half. He wanted to explain the field as currently understood and wrote to colleagues such as Bragg, asking for their latest material.[11] By late April, he was 'working like a slave over proofs & writing up papers.'[12] He also engaged in a vigorous correspondence with Boltwood, exchanging ideas: to Rutherford, Boltwood was in many ways part of the team.[13] When the new edition appeared in 1906 Rutherford made sure Kelvin was sent a copy. The great man was 'delighted to have it.'[14]

In the middle of this activity Rutherford got a job offer from Yale, which he declined. But he did find the time to travel to New Zealand, where the *Auckland Star* received him as a 'distinguished figure in the world of contemporary science' who had 'earned fame for himself — at an early age — which few men have ever attained,' work that meant his name would 'go down in history inseparably associated with that scientific marvel of the century, radium.'[15] The prodigal had, indeed, come home, but Rutherford was more interested in private time with his family. He spent time at Pungarehu and holidayed with family in the North Island's thermal district, which was becoming a new centre for tourism. He did, however, give one interview to the reporter from the *Auckland Star*. Once again he outlined the science then at the top of his mind, this time focusing on uranium as the original parent of radium and the time required to generate the proportions observed, which he expected to be at least a billion years.[16]

This was the first public mention of the method Rutherford envisaged for discovering the age of the Earth, and of the magnitude he had in mind, which was well in excess of earlier estimates. Kelvin, using heat dissipation alone, thought Earth at most 400 million years old. Rutherford knew that Charles G. Darwin and John Joly had been pondering the effects of radioactive heat. But Rutherford felt his disintegration theory, with its well-defined concept of half-life, offered more robust answers. Later in the year he began experimenting with ways of determining the age of rocks by analysing the proportions of lead and uranium, working with Boltwood to refine a technique. Other work at the time was producing values in the hundreds of millions of years, but Rutherford eventually concluded that Earth had to be around 3.4 billion years old. We know now that Rutherford's was the only figure of the day in the right general magnitude. The currently accepted age of Earth, around 4.55 billion years, was first defined in 1953 using a mass spectrometer to measure the decay products — a technique not invented when Rutherford came up with his figures — and has been only slightly refined since, including via moon rocks returned in 1972 by Apollo 17. Today the basic method for defining the age of any rock, now known as radiometry, broadly remains the one Rutherford helped pioneer.

Rutherford himself seems to have looked on the question of Earth's age only in passing, a side matter that emerged from his greater work on radioactive decay. His main focus of the day was determining, by experiment, the nature of alpha radiation. This also provided further data to counter criticism. And there was plenty of that. While Rutherford's colleagues in Britain applauded his work, the idea of atomic radiation by disintegration remained heretical and weird to others, including chemists. By this time Rutherford had a clear vision of atoms and was intending to explore their structure, which would provide final proofs of his concept. The specifics of alpha particles were a key element of that mix.

Not everybody accepted Rutherford's ideas. Chemists, brought up on the concept that atoms were solid and came together to form compounds, found his explanation of radioactivity to be simply wrong. From this perspective, Rutherford's argument that radium atoms included atoms of helium and lead meant that radium was a compound. This was a purely chemical interpretation that missed the basic concept Rutherford was presenting. He eventually vented his frustrations to Boltwood: 'These dam'd fools ... haven't the faintest notion that the disintegration theory has as much evidence in support of it as the Kinetic Theory of Gauss and a jolly sight more than the electromagnetic theory, which they all swallow as the eternal verities'.[17] His irritation at the scepticism with which his ideas were being met was perhaps reflected in the bold title he chose for the Silliman volume: *Radioactive Transformation*.[18]

This did not settle debate, and in the late summer of 1906 a vigorous correspondence erupted in the *Times* between some of the physicists.[19] Rutherford eventually wrote to *Nature* to make his own position clear.[20]

One of the main questions confronting Rutherford by this time was the source of radioactive energy. It was clear to him that the process of radioactive decay was enormously energetic by comparison with any chemical reaction. In the second edition of his book on radioactivity in 1905, and again in a letter to *Nature* the following year, Rutherford estimated the difference between chemical reaction and the atomic breakdown he had observed was in the order of a million.[21] Perhaps the more crucial point was the one he made clear in his last joint paper with Soddy, 'Radioactive Change' of 1903. While radioactive elements were unstable and able to release energy naturally, it was reasonable to suppose every element had the same potential. In a brief throw-away line, he and Soddy added that this offered a plausible way of explaining how the sun shone.[22] Once again it is clear that Rutherford had a vision in his mind of the wider meaning for physics, even though he had no explanations just then. And once again that gave him direction: later he discovered that he could, indeed, artificially induce a nuclear reaction in stable material.

The question, of course, was where that energy came from. The answer eventually emerged from theoretical developments that came in parallel with Rutherford's researches, work that provided some of the other key pillars of modern physics. Ultimately the directions that emerged also provided the explanations for Rutherford's atomic structure.

In hindsight none of this was surprising. By the turn of the twentieth century the fields of both experimental and theoretical physics were buzzing with activity as physicists picked up on the discoveries of the 1890s. As far as the theoreticians were concerned it was becoming clear that even electrodynamics did not offer the whole answer. One fundamental question remained the aether, which didn't appear to exist, except it *had* to exist in order to transmit electromagnetic energies. A paradox. There was also the problem of the photoelectric effect, which didn't manifest itself as classical physics predicted. Then an even larger theoretical challenge erupted. A little later the Austrian physicist Paul Ehrenfest (1880–1933) gave this issue the dramatic but slightly misleading name 'ultraviolet catastrophe'. It wasn't, strictly speaking, but the outcome certainly had plenty of drama about it, and the answers — as it turned out — led directly to a whole new set of concepts, ideas that eventually became the all-encompassing explanation for Rutherford's work: quantum mechanics.

The problem was a consequence of a phenomenon known as black-body radiation. This term was invented by Gustav Kirchhoff in 1860 as a way of wrapping science around the fact that any object that has absorbed electromagnetic energy will then radiate it. As an everyday example, any domestic kettle delivers energy into water, which is why the water eventually boils. Once the kettle is switched off the water immediately begins radiating the energy it has absorbed, in this case at infrared wavelengths, which humans feel as heat. This is why a forgotten cup of tea becomes cold. The mechanisms were identified in the seventeenth century as part of thermodynamics, but it was not until the mid-nineteenth century that physicists realised the phenomenon was electromagnetic and applied across the entire spectrum. The temperature at which radiated energy becomes visible light — starting with dull red — was defined in 1847 by the English chemist John Draper and became known as the Draper point.[23]

This idea became more complex when Kirchoff decided to wrap

CHEMISTS VERSUS PHYSICISTS

PERHAPS THE MOST public issue Rutherford faced with his concept of natural radioactive decay was the confusion it produced among chemists. During his first years at McGill, Rutherford suspected that the alpha radiation emitted by radium was an electrically charged helium atom — though he hadn't yet proved it — which to chemists meant that any material emitting it had to be a chemical compound of which helium was an ingredient. This was true in classical terms, where atoms were considered indivisible and a compound was produced with two or more different atoms. Water, for example, comprised two hydrogen atoms and one oxygen. However, Rutherford insisted that his process of radioactive decay was not chemical but a result of a structural breakdown within the individual atoms themselves: an atom of one element could split apart and produce atoms of one or more other elements. And that was where things tripped up. At this stage he didn't know the structure of the atom, nor all its components, so he could not detail exactly what was going on.

This question was why Rutherford then pushed on to investigate atomic structure and then turned his attention to the nucleus. He eventually discovered one of its component particles, the proton, which he identified in 1917–18, and then predicted the neutron. His colleague James Chadwick confirmed the neutron's existence on Rutherford's direction in 1932. With that information the mechanisms Rutherford had identified years earlier became explicable: the nucleus was self-fracturing and spitting out components such as alpha particles, which turned out to be the nucleii of helium atoms. In short, Rutherford was criticised by chemists, but then came up with answers about atomic structure that reset the entire basis on which chemistry worked.

Subsequent work has refined precisely what happens. As an example, radium-226, the most common isotope of radium, has 88 protons and 138 neutrons in its nucleus, each weighing about the same and giving it an atomic mass number of 226 (i.e. 88 + 138). The alpha particles emitted by an atom of radium-226 each consist of two neutrons and two protons — which happen to be a helium 4 nucleus — leaving behind 86 protons and 136 neutrons. The fact that there are different numbers of protons means this decay product is a different element, radon-222 (i.e. 86 + 136), which is a gas. This process of self-transmutation by radiation emission continues until a stable material is reached, typically a lead isotope.

ATOMIC DECAY

THIS DIAGRAM SHOWS alpha decay in action, and explains the direction of Rutherford's researches. The identification of alpha particles led to questions about the structure of the atom. The fact of a nucleus explained what he was observing in radioactive decay, but once he had the structure of the atom he wanted to know its composition, leading him to the discovery of the proton and prediction of the neutron. With that came a fuller explanation for radioactivity.

This diagram also shows how Rutherford's work was confused by chemists. Radioactive decay, in this case of a radium-226 nucleus, produced an alpha particle and a daughter product. The alpha particle was a helium-4 nucleus and the daughter product a radon-222 nucleus, but these were a result of nuclear disintegration, not because radium-226 was a chemical compound of the other two elements.

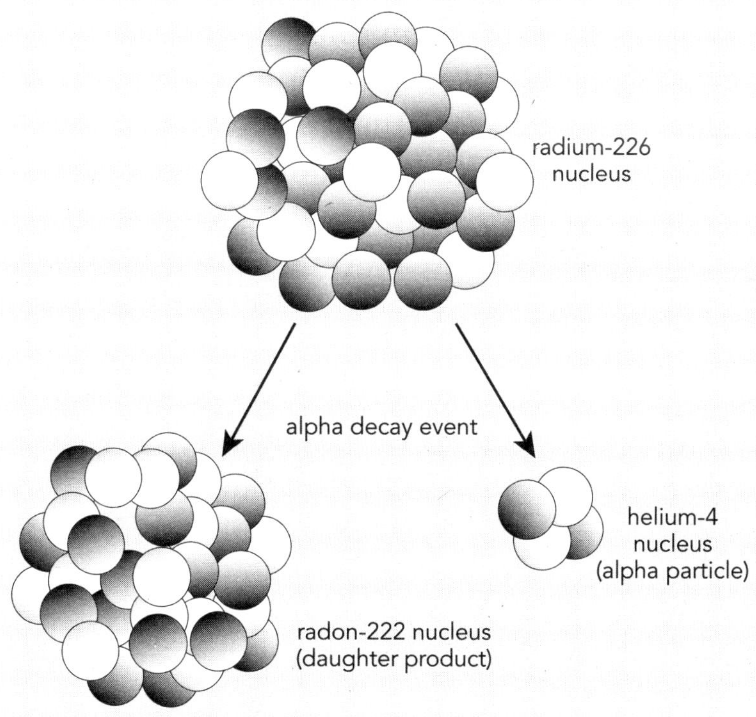

some mathematics around it by postulating what he called a 'black body', a perfectly non-reflecting object that absorbs all energy falling on it. This also made it a perfect *emitter* because the energy coming out of it wasn't mixed up with any reflections. His logic was similar to the old joke about physicists measuring the volume of a cow by first assuming the cow to be spherical.[24] Yet this general process of reduction was actually how physicists analysed phenomena then and, indeed, still do today. The rate of expansion of the universe, for example, is typically calculated by first assuming that all the mass and energy it contains is distributed evenly, something not true in reality.[25] One motive in this general approach is to simplify the mathematics. But in many cases the 'ideal' answer can also provide insight by contrasting that ideal with what is actually observed.

This last point is what Kirchoff had in mind with his concept of an ideal black body. Real materials weren't perfect emitters. However, Kirchoff's concept created an 'ideal' against which reality could be measured. This was given further shape in the mid-1860s by the Irish physicist John Tyndall (1820–93), who demonstrated a connection between the heat emission of platinum and the visible colour of its glow. The Austrian physicist and mathematician Josef Stefan (1835–93) added a mathematical treatment in the late 1870s, and the whole edifice was given a theoretical basis in 1884 by Ludwig Boltzmann, producing what is known as the 'Stefan-Boltzmann law'. This described electromagnetic radiation emitted by any object solely in terms of what was known as 'black-body temperature'. Today the colour of household lighting is often defined this way, for example a typical LED bulb is listed as 2700 degrees Kelvin, meaning the light it produces is 'warm white'. Colour temperature in this sense is also important in photography.

Boltzmann's mathematics were tested in 1888 by the German physicist Heinrich Weber (1843–1912) and found to be true. Then, in 1893, the German physicist Wilhelm Wien (1864–1928) devised what is now known as 'Wien's law', in which he showed that the quantity of energy re-radiated by a black body was inversely proportional to the temperature. This wrapped a physics principle around the fact that hotter objects emitted radiation at shorter wavelengths than cooler. At the time this was couched in terms of prevailing 'classical' physics — the principles devised by Newton and his colleagues, coupled with Euclidean geometry.[26]

All this became part of the emerging exploration of electrodynamics, the mathematics being developed to describe it, and the idea that electrodynamics was the fundamental principle behind physics. But then, around the turn of the twentieth century, the concept of black-body radiation tripped up. The difficulty came when Rayleigh attempted to show a relationship between the scale of emitted energy and its frequency. According to his mathematics the emission energies spiked to infinity at ultraviolet frequencies. This was clearly absurd, and nor was this matched by experimental data. But the calculations were correct for other wavelengths, so something had come adrift. Rayleigh published his findings in June 1900. By that time the German physicist Max Planck (1858–1947) had been working on the same issue for a while on the basis of experimental data. He was unaware of the British efforts and presented his explanation to the German Physical Society in December, later published as a ten-page paper in *Annalen der Physik* (*Physics Journal*).[27]

Planck's answer derived from a statistical analysis and made the assumption, radical for the day, that energy was radiated in discrete chunks or 'quanta'. This meant it stepped from one energy level to another instead of increasing smoothly. Once re-calculated on that basis, the black-body curve lost its nonsensical 'spike' at ultraviolet frequencies.[28] The mathematics worked out, but in classical terms the idea of stepped quantities seemed absurd. Planck couldn't explain *why* energy was released in discrete packets. He remained in doubt himself about his numbers and was not reassured until Rutherford later provided empirical proof, by experiment.[29]

The crucial import of Planck's idea was that it challenged classical principles. This was new. By this time the main underlying focus of both theoretical and experimental physics was the effort to re-frame everything in terms of Maxwell's electrodynamics, starting with the relationship between electromagnetic forces and matter. That meant overturning some of the conceits of Western science as it had evolved to that point, notably the idea that the elements were immutable. But the basic principles — Newton's mechanics and Euclid's geometry — remained the fundamental basis on which electrodynamics were thought to work. Efforts to understand how electrodynamics worked, including Maxwell's own paper, were still framed in terms of classical principles. Rutherford's working papers, for example, include equations that he developed, derived from both Newton's mechanics and Euclid's geometry.[30] This was understandable. The principles clearly worked — and worked well — in an everyday sense, so the focus was on figuring out *how* electrodynamics worked in those terms. And for the most part, that was also successful. But it was becoming clear that this wasn't the whole answer.

Rutherford's former mentor and now colleague J.J. Thomson had been working at the Cavendish on his own researches into atoms and subatomic particles, and in 1904 published a remarkable paper describing an electrodynamic structure for the atom. This was a watershed, though not because Thomson was right. While his atomic model went down well in Britain, his European colleagues were scornful. But the importance lay in the fact that he had made an effort to reconcile his observations with electrodynamics. He had shown experimentally that atoms contained electrons, which he persisted in calling 'corpuscles'. Now he proposed that atoms were made of them. His description became known as the 'plum pudding' model by analogy with the British dessert, though in fact Thomson explicitly described it as 'an electrodynamic soup' within which electrons orbited.

Thomson had no evidence other than the existence of the electron and experimental data associated with charges, but was able to show that his idea worked mathematically, devoting the first 16 pages of his paper — nearly half the publication — to equations. These described the way negatively-charged electrons could be organised in discrete orbits within a positively-charged 'soup'.[31] He also explained how atoms might lose electrons, which accounted for ions. The interesting point about Thomson's model is that it was the first to include a subatomic particle, the only one then known. It was framed wholly in electrodynamic terms, and it required electrons to have discrete energy levels. These were major breaks from earlier thinking. Thomson's model was superseded by Rutherford's, as we will explore in the next chapter — but the fact that Thomson made such a proposition at all was indicative of what was happening within the field.

Atomic structure was not the only mystery that needed explaining in terms of the new electrodynamic paradigm. The issue of missing aether nagged. So too did the photoelectric effect. And Planck's claims needed proving. Into this fervent mix came a 26-year-old physicist working at a patent office in Berne who, in a remarkable outpouring of papers in 1905, proposed bold answers to the key questions. His name was Albert Einstein, and his eventual impact was so profound that even today he remains a household name. The shock of unruly grey hair he cultivated in his later years still defines the pop culture image of a scientist, echoed in the look of Rick Sanchez from the *Rick and Morty* television series, among others.[32]

This was not bad for a young man from Ulm, Germany, who had bombed out of school early. Einstein was born in 1879 to Hermann and Pauline Einstein. His interest in the sciences was apparently triggered at age five by his observations of a compass, later by a book on geometry. Then, when Einstein was 15, his family moved to Italy, settling in Milan. Hermann Einstein had branched into engineering by this time and apparently hoped his son might follow suit. Albert was less enthusiastic. School was disastrous: the young Einstein simply didn't fit the period system of rote learning. But he taught himself mathematics, apparently inspired by a regular family guest, Max Talmey, who introduced him to Aaron Bernstein's popular science books. From Bernstein's books, and particularly a thought experiment in which the reader had to imagine travelling down an electric wire, Einstein wrote a paper on electromagnetism and aether, which he sent to his uncle, Casar Koch.[33]

Like Rutherford, Einstein did not initially know what he wanted to do. And — again in a curious parallel to Rutherford's youth — his father's businesses went through wild oscillations. For Einstein the more crucial issue was that all young German men were required to do military service. Einstein dodged the draft and went to Switzerland, where he tried to get into the Swiss Federal Technical School. He lacked the entrance qualification and failed some of the entrance examinations, but was granted entry providing he completed his basic schooling. He graduated in 1900 but failed to get work as a teacher. He became a Swiss citizen the following year and ended up working in the Zurich patent office, a scientific 'civvie street'. Here he moonlighted as a physicist, spending his evenings wondering about the nature of light and working on his doctoral thesis.[34]

This work came together in 1905 when Einstein submitted his thesis and published four groundbreaking papers.[35] All were expressions of the same broad ideas, driven by what he later described to the Dutch physicist Willem de Sitter (1872–1934) as *Verallgemeinerungsbedürfnis* ('need for generalisation'). Together, Einstein's 1905 papers presented a startling vision that, like Rutherford's concept of atomic transmutation, was heretical by classical standards, but also established new paradigms along with a call for further research.

Einstein's 'annus mirabilis' (miracle year) began in March when he released a paper on the photoelectric effect, drawing from Planck's concept of energy steps. Maxwell had proposed that electromagnetic forces such as light were waves. However, Einstein argued, in the material world with its constituent atoms, energy by definition had to be stepped because each atom or electron represented a specific quantity of energy. So what would happen, Einstein wondered, if light — meaning electromagnetic energy — was not a wave, but a stream of particles? When he analysed the problem he realised this could explain the photoelectric effect. The equation Einstein devised to describe this relied on Planck's hypothesis and was a key preliminary step towards quantum mechanics.

At the time the reason why Einstein's proposal was so important was that it did away with the aether. If electromagnetism was a stream of particles then it self-carried and there was no need to keep looking for an invisible and undetectable material to do the job. It was a significant breakthrough, albeit one that was not fully recognised for some time: Einstein had no empirical proof, and his assertion didn't account for the fact that electromagnetism still showed wave-like properties. It took more than a decade to sort out the answers, but in the end Einstein was shown to be right and received the Nobel Prize for his photoelectric explanation in 1921. But this did not answer the question: how could electromagnetic energy be *both* a wave and a particle?

Einstein's photoelectric paper was remarkable enough, but it was only the beginning of an extraordinary year. At the end of April 1905 Einstein submitted his doctoral thesis, using the size of sugar molecules to derive Avogadro's number — something crucial to both chemists and physicists because it defined the number of particles in a given volume of gas. This was a value that Rutherford later investigated himself. Incredibly, Einstein's thesis was less remarkable than a paper he issued that same month on Brownian motion. This is the random vibrational movement of particles in response to energy — for example, the movement of water molecules in a hot cup of coffee. Einstein provided an explanation for it that redefined the kinetic theory of heat, which Rutherford had likely used earlier in his 1900-era estimates of the energies involved in particle experiments.

Einstein's final two papers of the year took an even larger step, one that ultimately explained some of Rutherford's discoveries and showed how entwined they were with the basic frameworks of physics. The title of Einstein's larger paper, 'Zur Elektrodynamik bewegter Körper' ('On the electrodynamics of moving bodies'), extended electrodynamics to the geometry of space. Up to that point Western scientists had broadly framed their thinking around the geometry Euclid had defined more than two millennia before and which Newton then wrapped with physics. That came into question once electrodynamics entered the picture, and Einstein now threw a further wrench into the works. To do this he built on the equations of Maxwell and Hertz, into which he mixed the work of Lorentz and FitzGerald.[36] These scientists had explained Michelson's failure to detect aether drag by arguing that space deformed like an electromagnetic field. This was weird in everyday terms, but Einstein took the idea and ran with it, leveraging the fact that every measurement of the speed of light — or the speed of electromagnetic radiation — came up with the same value. Maxwell's equations were based on the idea that light-speed was constant. To Einstein that carried some specific implications.

To illustrate the point, imagine Edmond Halley riding on a horse-drawn carriage. His friend Isaac Newton stands nearby on the ground and measures the speed of the carriage at 10 km/h. Both scientists then throw a tennis ball forwards in parallel directions at 15 km/h. Newton measures the velocity of their respective balls and finds that his is moving as expected at 15 km/h, but Halley's is moving at 25 km/h. This is obviously because Halley is already moving at 10 km/h and the velocities have been added together. Now imagine an astronaut inside a spaceship moving at light-speed past Einstein. The astronaut shines a torch out the front, sending a beam of light forwards. Einstein, standing still as the spaceship passes him, does the same, paralleling the astronaut's beam of light. Then he measures the velocity of both. Is the astronaut's beam moving at twice the rate of Einstein's, by analogy with Newton's measurement of Halley's tennis ball? No. By contrast with Newton, Einstein finds that both beams are moving at the same velocity. In other words, the speed of light is constant.

This idea carried profound implications. Newton had shown that velocity, distance and time were directly related: a vehicle that moved a distance of one kilometre in one hour was obviously moving at one kilometre per hour. In this vision, thanks to Euclid's geometry, distance was fixed and the other two were directly related variables: the same cart moving at twice the speed could cover the same distance in half the time. What happens, Einstein wondered, if the fixed value is velocity? In that case, Einstein reasoned, the other values had to be variable. That matched FitzGerald's hypothesis about length contraction. But it was a radical vision. All depended on the relative position of observers — technically, their 'inertial reference frame'. This produced the name eventually applied to the theory: 'special relativity'. The mathematics also pointed to another conclusion: the speed of light is a hard limit. Anything that has mass *cannot* travel at the speed of light, and anything without mass *must* travel at the speed of light.

In some ways Einstein's papers on the photoelectric effect and special relativity were flip-sides of the same coin. In one he envisaged what would happen to Maxwell's equations if electromagnetism was transmitted in discrete particles, and in the other he envisaged what would happen if space, time and mass obeyed the rules of an electromagnetic field. Together they also produced a truly bizarre notion for the day: light was carried by a stream of particles, but these particles couldn't have any mass, because they travelled at light-speed. But if they didn't have any mass, how could they exist as particles?

Einstein's answer came in his fourth and final paper of 1905, which was a short piece and essentially a technical note: 'Does the inertia of a body depend on its energy content?'[37] Einstein was not the first to consider inertial energy and electrodynamics — it had been on Rutherford's mind as a result of his work on radioactivity, and others had also been looking into it. However, Einstein's relativity added a twist. He was able to show mathematically that, in a universe of constant light-speed and flexible space-time, mass and energy were different aspects of the same thing. Again, this was not as radical as it sounded — things could not be otherwise if matter was to obey Maxwell's equations.

There has been some question since as to whether Einstein's concept of mass-energy was original. The equation always associated with his name wasn't in either of his papers on special relativity in that form, though he was later credited with it by his colleagues.[38] However, he was the first to provide a mathematical derivation based on special relativity, where light-speed was a constant. The actual equation Einstein produced to describe it was complex because the conversion had to take motion into account. What is also forgotten are the words with which Einstein ended his note, a call to test his ideas with high-energy materials such as radium.[39] He had no data to back his hypothesis and the effects he was describing were logarithmic, obvious only at extremes of energy and relative velocities. Detecting this in the everyday world was going to be difficult, which is partly why Newton and his colleagues never discovered the principle.[40]

Today we know that Einstein was right: systems to compensate for the time-dilation aspect of special relativity even have to be built into GPS satellites in order for them to be accurate, meaning that anyone using a smartphone can demonstrate Einstein right, multiple times daily. But in 1905 there was no evidence, and Einstein was clearly wondering whether the extreme conditions associated with radium emission might provide data. As it turned out, his principle of mass-energy equivalence explained the energies Rutherford was finding in radioactive materials, though it wasn't as direct as a simple mass-energy conversion of whole atoms. The key to it was Rutherford's discovery of the nucleus, but even then physicists took some decades to find the mechanisms.

All these developments came thick and fast in just a few heady years from the turn of the century and made clear that a science revolution of the kind identified by Thomas Kuhn was well under way. Physicists had responded to the late-nineteenth-century crisis. New ideas were being explored and the main foundations of what we could call modern physics were taking shape, including the first concepts of quantum mechanics and the nature of particle physics. All were driven by an effort to reconcile the concepts of electrodynamics first with classical physics and then — when that didn't work —with the new discoveries opened up by nineteenth century technology, beginning with radioactivity.

This new approach also showed that strictly classical thinking wasn't going to cut it. Newton and his colleagues hadn't been wrong in terms of what they knew, and the basics they had identified still

EINSTEIN'S EQUATION AND RUTHERFORD'S RADIOACTIVE ENERGY

THE RELATIONSHIP Einstein proposed between mass and energy has become the symbol of geeky pop culture: $e = mc^2$. The conversion factor was the square of light-speed, which Einstein represented by the letter c, meaning 'constant'. It meant that a very small quantity of mass was the equivalent of a gargantuan amount of energy. Imagine a marshmallow with a mass of 7 grams. If wholly converted to energy this could run a household 2-kilowatt fan heater for 87.3 *million* hours, around 9,965 years.[41] This is just over four times the total electrical energy generated in Rutherford's home country of New Zealand in the calendar year 2023.[42]

The problem with such a scenario is that Einstein never suggested mass and energy were interchangeable at will: the relationship was an outcome of relative velocity in a universe where light-speed was a constant. That was why the actual mass-energy equation in his 1905 note included motion. Later, the discovery of particles with an opposite electric charge — detected by Rutherford's colleague Patrick Blackett in 1932 — showed it was actually possible to convert mass to energy by colliding matter with antimatter, causing the electric charges to cancel each other out and releasing the entire energy equivalent of the masses involved. But this did not happen with ordinary matter, and it was later discovered that antimatter forms a tiny proportion of the mass-energy of the universe, an asymmetry that continues to puzzle physicists today.

Einstein's principle of mass-energy equivalence nonetheless explained the energies being found by Rutherford and others in radioactive material. It took time to discover exactly how, but the key to it was Rutherford's discovery of the nucleus and subsequent push to identify the particles within it. He knew something was going on within the nucleus that wasn't fully explained by electromagnetism, an issue made all the more obvious by his discovery of protons inside the nucleus. These were positively charged, meaning they should repel each other: in other words, a nucleus shouldn't be possible in electrodynamic terms alone. The specific force involved was discovered in 1935 when the Japanese physicist Hideki Yukawa (1907–1981) published an analysis of the interactions between protons and neutrons. Even then it was the mid-1960s before the full answer began to emerge. It turned out that a force existed within the nucleus that bound together the particles comprising protons and neutrons, and which also bound the nucleus, within certain limits. This became known, perhaps predictably, as the 'strong force', and today has been shown to be about 100 times more powerful than electromagnetism. Its discovery laid the groundwork for a full explanation of the energies involved in nuclear reactions, and why elements heavier than lead always decay to lead.

As currently understood, strong force grips the protons and neutrons in the nucleus with such intensity that the net mass of a nucleus is less than that of its components in isolation. When two nucleii crash together and fuse, the difference is released as energy, in quantities defined by Einstein's equation.[43] This is why the sun shines: specifically by fusing hydrogen into helium in its core and releasing energy in the process. The quantity of energy involved is equivalent to about 0.7 percent the mass of the hydrogen that is fused — in other words, it is far from a total matter-energy transformation. But the quantities of energy produced by the Sun are nonetheless colossal because the Sun, itself, is enormous.

This effect is not consistent across the periodic table, primarily because strong force is very short-ranged. It binds together the particles that form protons and neutrons, but the intensity falls off rapidly outside that radius. This means that while strong force can overcome electromagnetic repulsion forces in a smaller atomic nucleus, electromagnetic repulsion has increasing effect in larger nuclei.

This tension between strong force and electromagnetic repulsion also limits the weight of stable elements. The heaviest is lead, which has four stable isotopes, each with 82 protons.

The nuclei of heavier elements break down at various rates until they are reduced to lead. This breakdown is the main origin of alpha particles and the decay chains Rutherford and his colleagues mapped out. It remains a sign of Rutherford's genius that he was able to identify that radioactivity was due to breakdown at atomic level, even before he identified the nuclear structure of atoms.

The same limit to strong force effectiveness also explains why the heaviest naturally occurring element is uranium-238, the 92nd element in the periodic table. Heavier elements can can be artificially made in cyclotrons or particle accelerators, but are increasingly short-lived as the atomic weight increases. For example, Rutherfordium-265, the 104th element, was initially synthesised by Soviet physicists in 1964 and includes isotopes with a half-life of one minute. Organesson-294, synthesised in 2002 as the 118th element, has a half-life of only 0.7 milliseconds.

The energy released when a nucleus breaks down is again defined by Einstein's mass-energy equation. It turns out that the total mass of the products of a decay event is slightly less than that of the original atom, a difference made up by energy released in the process. This is far less than the energies associated with nuclear fusion but still enormous by everyday standards. Furthermore, strong force mechanisms are not the only ones going on in an atomic nucleus. A different short-range force, known as the weak force, causes neutrons to break down into a proton, an electron and an antineutrino. This is the cause of beta radiation.

These understandings were developed during the mid-to-late twentieth century following Rutherford's discovery of the nucleus. He did not have these later details, but knew the radioactive materials he worked with were incredibly energetic as a result of breakdown, estimating the energy involved was up to a million times that of an equivalent mass in a chemical reaction. Studies have since shown this estimate to be essentially correct.

applied. But the new physics showed that something else lay behind what they had observed. And some of that was going to overthrow classical ideas completely. Rutherford was essentially first off the block with his explanation of radioactivity as atomic disintegration, setting up a starting point for both nuclear science and the wider field of particle physics — the basic building blocks of matter and energy. Planck provided the first concepts of what we now call quantum mechanics, which eventually supplanted classical physics at subatomic levels. Einstein added to that picture by recasting both Euclid's geometry and Newton's mechanics around the principles of Maxwell's electrodynamics and the concept of light-speed as a constant.

Bringing all this together involved a great deal of work, guiding the direction of physics during the first three decades of the twentieth century and creating the idea-set still in use today. The direction in 1905 seemed clear enough to those involved. Einstein decided to find out whether the relativity he had defined could be extended to a general rule, which meant he had to find a new theory of gravity. Rutherford, meanwhile, pressed on with his investigations into particle physics, in particular trying to discover the nature of alpha particles, which he expected would give him insight into the structure of atoms. So too did the Curies in Paris, among others.

In many respects it was no coincidence that all this emerged as the twentieth century turned. Western culture was in the ascendant. The complex societies and technology it built enabled the physics community to flourish across the Western world, its colonies, and to some extent into other cultures that were directly engaged with that world: Japan and India in particular. In the case of India the engagement was because the subcontinent was part of the British Empire. Japan, meanwhile, was actively engaging with Britain and was an ally from 1902. Japanese physicists spent time in British establishments.

The discoveries that followed were extraordinary by any measure. But where the edifice fell down was that the underlying frameworks of thought were still built around the styles of thinking that had emerged in the seventeenth century, with their particular frameworks. The fact that Rutherford, Einstein and Planck had disposed of key concepts of earlier physics paradigms, or that their work bridged the arbitrary divisions between 'different' sciences, did not mean they had also thrown out scientific method. Observations were still followed by an effort to find a pattern, and a hypothesis then devised to explain that pattern, just as Newton, Hooke and others had done 250-odd years earlier.

The detail was subtly differentiated across Europe's major nations: there were variations between approaches by physicists from Germany, Britain and France, the main powerhouses of the field at the turn of the twentieth century. And then there was Rutherford, whose mindset was shaped by his upbringing in New Zealand. But at the broadest level their approaches matched the way this kind of thinking had been expressed across the span of Western history since the sixteenth century.

One of the outcomes was that the new vision was framed by this thinking, including the way that science categorised its observations. This ultimately led to a series of contradictions. Some were already clear from Einstein's paper on the photoelectric effect, renewing the long-standing puzzle over whether light should be categorised as a wave or a particle. These were totally different things to Western thinking. But this was a product of the way Western science looked at the phenomena, not the phenomena itself. Europe's new science, in short, was revealing the deeper secrets of the universe but then stumbling over its own determinism.

8 | Journey to the centre of the atom

Rutherford's ability to seize on one definite fact, to realise that independently of all other arguments it showed a fundamental error in the current physical picture, and to then follow this trail ... was the measure of his genius.

— Charles Drummond Ellis (1895–1980)[1]

IN OCTOBER 1906 RUTHERFORD discovered that he had become an integral part of geology. 'Marckwald has named a mineral after me,' he declared to Hahn.' '"Rutherfordin"; ye gods!'[2] To have his name put to uranyl carbonate — formally spelt Rutherfordine — which had been just identified by German chemist Willy Marckwald from samples found in what was then Tanganyika,[3] was a salutary endorsement. But Rutherford had no intention of resting on his laurels. The way ahead was obvious, and he had already laid it out in the revised version of his 1904 book. Physicists, he declared, were focusing their attentions on the relationships between electricity and matter. Studying radioactivity had helped provide insight, showing that the 'atom itself is not the smallest unit of matter, but is a complicated structure made up of a number of smaller bodies.'[4] He was determined to find out just what that structure might be — something likely to go a long way towards explaining why some elements were unstable.

This led to what is often listed as Rutherford's second great breakthrough, his discovery of the atomic nucleus. And this was a significant breakthrough by any measure, although from Rutherford's perspective it was not due to any new interest, but simply a further step in his consistent exploration of radioactivity and its causes. He did not make that discovery in Canada. At McGill, Rutherford had one of the best physics laboratories in the world. The university wanted him and kept upping his salary whenever he seemed to be tempted by work elsewhere. But offers kept coming, and by this time Rutherford felt he had reason to move on. In Canada he was away from the epicentre of physics, and he missed the discussions he had enjoyed at the Cavendish. In mid-1906 he took up an offer to visit the University of California at Berkeley, where — as he told Hahn — he 'had a very good time', lecturing to 'a class of 30 to 40, mostly professors and instructors.'[5] The campus was in good shape, but nearby San Francisco was still in ruins after a disastrous earthquake in April, to Rutherford a 'most depressing sight.'[6]

Around this time Yale shoulder-tapped him with an offer, but Rutherford decided to stay where he was, pending an opportunity in Britain where he would 'not have to sacrifice laboratory facilities.'[7] He discussed prospects with Arthur Schuster, then holding the Langworthy Chair of Physics at the Victoria University of Manchester, who was thinking of retiring. Rutherford did not think anything would emerge any time soon — he had even bought land near Montreal for a house — but then in September he got a surprise letter from Schuster, who was on his way and wanted to confirm Rutherford's interest. Schuster hinted at a salary of £1000 with possible increases to £1200.[8]

Rutherford was cautiously interested. As he told Schuster, the laboratory was a 'great attraction', as was the 'opportunity of more scientific intercourse than occurs here.'[9] But he did not want to lecture more than he had to — five lectures a week 'but no more.' Nor did he want to get heavily involved in committee work.[10] This opened what amounted to negotiation for terms and conditions whereby Rutherford was trying to maximise his research time, to which Schuster responded with what amounted to sales pitches, telling Rutherford on 7 October that he been able to hire the Scottish physicist James Dewar's instrument maker who 'works by agreement on cheap terms for us and is an excellent man.'[11]

Finally, on 8 November, Schuster cabled Rutherford to tell him he had the job,[12] and Rutherford was formally offered it in December.[13] He formally resigned from McGill on 1 January 1907.[14] The physics grapevine was shortly abuzz with the news. Soddy, by this time at the University of Glasgow, alerted Bragg just days after Rutherford was appointed. 'It will make a great difference having him over here & he should have a fine opportunity at Manchester which is very "progressive" & free from traditions.'[15] Edward Nichols, then professor of physics at Cornell University, congratulated Rutherford with the rider that 'We are all of us very sorry to lose you from our shore and shall miss you.'[16] He was still lamenting Rutherford's departure months later. 'I'm still very sorry you went so far.'[17] Rutherford was also sorry to leave McGill, but he saw many advantages to the move, pointing out to Principal Sir William Peterson that it put him at the epicentre of the physics community.[18]

The move to Manchester was sensible on multiple levels. Rutherford had control of one of the best laboratories in Britain and had been able to negotiate access on terms that favoured research and funding. He was in the middle of a vigorous physics community, meaning he had a better selection of graduates and colleagues to work with. The department administered three major scholarships — the John Harling Research Fellowship, Heginbottom Scholarship and Hatfield Scholarship — which offered money to worthy students. Of these, the John Harling Research Fellowship was the most important from Rutherford's perspective because it paid for a full-time researcher.

There may also have been another advantage. Despite every precaution McGill's laboratory was becoming increasingly contaminated, with dire effects on Rutherford's instruments. He admitted to Boltwood late in 1906 that his premises were 'in a perilous state for fine experiments.' He knew just how enduring some radioactive decay products were and saw 'no hope of improvement.'[19] The practical answer was to swap that laboratory for another — something that was likely a factor in Rutherford's decision.

The physics laboratory at Victoria University of Manchester was set up in 1870–71 by Balfour Stewart, the author of Rutherford's childhood physics text. It grew from eight students in its first year to 183 in 1895–96.[20] Schuster was appointed professor in 1888 and helped transform the laboratory into one of Britain's leading research centres, raising around £40,000 to support it. In the late 1890s he persuaded the university to fund a new building. Work on it began in October 1898 and it was formally opened by Lord Rayleigh in June 1900. Facilities included lecture rooms on the second floor, laboratories, a workshop, and even a small telescope on the roof.[21] By the time Rutherford arrived there were about 150 undergraduates and another 100 on postgraduate and other work.[22]

So Rutherford left McGill after almost a decade during which he had done a great deal to establish the repute of the physics department, swept like a whirlwind through the medical departments in his efforts to support radiotherapy for cancer, accidentally discovered the principle behind smoke detectors, and helped put McGill on the world map of the sciences. The university awarded him an honorary higher doctorate — LLD — in thanks for his work. Cox, his long-standing boss, went a step further and nominated Rutherford for the Nobel Prize in recognition of his work on radioactivity. The application came too late for the 1907 prize but was held for the following year. Rutherford's place in McGill's history has since been memorialised in the university's physics building and a sports park on Pine Avenue.[23] His equipment also survived. As the physicist F.R. Terroux remarked in 1938, it would ordinarily have possessed 'neither intrinsic nor obvious aesthetic value' and would normally have been plundered for parts. But Eve, who became Rutherford's successor, made sure the equipment was preserved.[24]

Rutherford's last months in Canada were frenetic. There was a series of public lectures, along with a dinner in honour of Joseph Larmor in New York.[25] He had work to finish, correspondence to deal with — including the usual array of public queries.[26] By June Rutherford and his family were in England making arrangements to settle in Manchester, ready for the new university year that was due to start in October. It was a busy time. He listed some tasks in a letter to his mother: he had been to a meeting with the Royal Society, attended their soiree at Burlington House, and then met the Chemical Society.[27] He was, in short, back in Britain's scientific society. He found a place to live in Withington, not far from his laboratory. In the middle of all this, as he told Bragg, he also 'found time to begin my investigations' on the effects of temperature on radium.[28]

That work highlighted one of Rutherford's initial challenges at Manchester: getting a good supply of radionuclides, especially radium. The various elements Rutherford had used at McGill remained there, although it appears he did take some of the actinium.[29] There were also supplies at Manchester: Schuster had obtained radium for his own work and that of his students, which Rutherford inherited. But he needed more, and he told Boltwood he 'hoped to raise some' by year's end, adding: 'This is one of the advantages of being on the fighting ground instead of at a lonely outpost.'[30] Part of the issue was cost: radio-elements were expensive. In July, a few weeks after arriving in Manchester, Rutherford arranged to borrow around 40 kg of uranium residue owned by the Royal Society, from which he thought he could extract actinium[31], sending some to Boltwood and asking for advice as to the best method. 'I shall probably have a chemist or two to turn on to it.'[32] He began experiments with the residues in October.[33]

Rutherford also approached the *Kaiserliche Akademie der Wissenschaften* of Vienna ('Vienna Academy') in hope of borrowing radium.[34] They had what was arguably the best supply in the world, but were prepared to lend it only if Rutherford shared the stock with William Ramsay, who was still at University College. However, when Ramsay received the delivery he refused to divide it, declaring the radium 'infinitely more valuable as a whole' and offering to hand it over 'after a year or a year and a half.'[35] Rutherford reluctantly agreed to what amounted to an ultimatum, as long as a 'definite arrangement' could be made.[36] What followed was awkward: in what seemed to be an apology for his refusal to split the batch, Ramsay offered Rutherford the 'emanations' from it, but then lost the first amount he had collected and went on holiday.[37] Rutherford gave up, went back to the academy and arranged to directly borrow another 500 mg of radium bromide. He reported it to Boltwood with a certain smugness: 'I am now quite independent of Ramsay and have got more than he has — so should selfishness (not of my self) always be rewarded.'[38]

Having a fair quantity of radium made it possible for Rutherford to examine the 'effects produced by the radiations' on a 'comparatively large scale.' He found ways of maximising the effects with what he called 'alpha ray tubes', narrow glass straws that channelled the alpha particles.[39] The improved radium resource and switch of research focus to this material, by comparison with his McGill work, is given dimension by the fact that many of Rutherford's notes from 1909–1910 became so radioactive that today they carry warning labels and require special handling.[40]

Rutherford collected a team of graduates to support his work. At McGill he had looked to chemists for assistance. He did so again as he assembled his Manchester team, including the Polish chemist Kazimierz Fajans (1887–1975). Late in 1906, as he contemplated how he might shape up his position in Manchester, he decided he needed support in 'mathematical physics', asking Schuster if he might take on that role.[41] The point makes clear that Rutherford knew where his research was going. His students were handed what Rutherford considered to be the dull parts, but he also leaned on specialist expertise. One of the newcomers was a young German doctoral graduate, Johannes ('Hans') Geiger (1882–1945), who had a John Harling Fellowship to study at the university. Rutherford recognised Geiger's talents and worked with him to solve some of the key questions about alpha particles. Rutherford also treated his team like family, bringing them together at his home for Sunday dinners. There were reasons why he gained the nickname 'Papa'. Only Geiger didn't use it, referring to Rutherford instead as 'the Prof'.[42]

A return to Britain put Rutherford within easy distance of friends and colleagues, and he lost no time renewing his connections, opening his home to visitors.[43] His networks both in Britain and on the continent expanded for these reasons, and he made a point of keeping up with colleagues after they had left his laboratory: when Fajans left the Manchester laboratory for a post in Germany in 1910, for instance, Rutherford kept up a steady correspondence with him, variously in German and English, for years.[44] By the same token, he was now more isolated from his American friends, keeping in contact by mail — though it was not always two-way. 'I am not really dead — only a bad correspondent,' Henry Bumstead wrote in June 1908. 'My laziness about writing is encouraged since I hear from you through Boltwood.'[45] Rutherford particularly missed Boltwood's wisdom and invited him to Manchester for a year in the laboratory. Boltwood initially declined.[46]

Rutherford's enthusiasm to keep up with all in his network provoked one clanger. Late in 1907 he was visited by the Principal of the University of Leeds and learned that the university intended to offer Bragg the position of professor of physics, just 60-odd kilometres from Manchester. He was overjoyed and posted a lengthy letter to Bragg in Adelaide.[47] Bragg responded promptly: 'You told me before I actually had the offer what the terms would be.'[48] Bragg accepted the job and moved back to Britain with his family, including his son William Lawrence Bragg, who became a physics student at Cambridge. The elder Bragg was responsible for introducing a new research method: X-ray crystallography, complementing the approaches Rutherford was taking at Manchester.

By the end of 1907, then, Rutherford had moved countries, moved universities and plunged back into the social world of British physics.[49] He was 36 years of age, still a stripling in a field where professors were meant to be white-haired elderly gentlemen. The point came out when he was introduced to Baron Kikuchi Dairoku, a leading Japanese academic who had been educated in Britain. It turned out that Kikuchi thought he had just met the son of the 'celebrated Professor Rutherford.'[50] The incident made clear just how far Rutherford had risen in the field. His standing was also recognised by his former teacher, Bickerton, who approached Rutherford wanting support for his mildly crackpot theory that two partially colliding stars might produce a third from the detritus. The theory had no basis even in period knowledge. Bickerton retired back to Britain in 1910 and continued to push the idea to Rutherford, who found himself having to gently manage this curious reversal of role.[51]

One of the issues with the radium bromide being bought or borrowed around the physics community was its variability in output or 'strength.' By this time a variety of colleges and universities were obtaining supplies, and Rutherford was occasionally approached to test radium against his own stock.[52] Both Marie Curie and the Academie des Sciences in Vienna had produced their own standard measures of radioactive output. Rutherford came up with one as well. In 1910 Rutherford got involved in an effort to establish a baseline standard across the field. This meant getting Curie to buy into it, as she had the best quality material. His draft letter to her has survived, filled with strike-outs and emendations: he was clearly struggling with how best to get her on board.[53]

What followed was driven more by politics than science. A major conference on radiology was due to be held in Brussels in September. Curie arrived, looking to Rutherford 'wan and tired and much older than her age.'[54] Rutherford and a number of colleagues — including Boltwood — sorted out what they wanted behind the scenes, then formed an International Radium Standards Committee to finalise arrangements. Decisions included an agreement to name the new unit

RADIUM SALTS

BY THE END OF THE nineteenth century Europe's chemists had a number of methods for refining radioactive ores such as pitchblende and isolating the elements they carried. The resulting material was vital for the research then being carried out across the Western world, initially because physicists and chemists were investigating these materials for their own sake; and later because the radioactive particles produced by radium, in particular, were highly energetic and provided a tool for research purposes.

Most of the radioactive materials were produced as salts. The Curies first isolated radium in the form of radium chloride. Later, radium was primarily extracted as a salt of bromine, radium bromide. This carried multiple issues. One was that batches were of differing strengths, as Rutherford discovered to his cost on arrival at McGill. However, the main problem facing researchers in the first years of the twentieth century was that a tonne of pitchblende typically contained only around 0.15 grams of radium. This made radium difficult and expensive to obtain, stretching physics departmental budgets. That didn't change quickly. As late as 1913 the world supply of radium was reportedly only about 8 grams, though production was ramping up as the supposed 'health' benefits were commercialised.

None of this was risk-free, as both Marie Curie and her daughter discovered to their cost. Radium is one of the most toxic materials ever discovered by humanity. The half-life of its longest-lived isotope, radium-226, is just under 1600 years, meaning that radium used by these physicists 120-plus years ago still retains about 95 percent of its original radioactivity today. The main hazard of radium is that it is an alkaline earth metal, in the same family as calcium. If a radium salt is ingested the body treats it as a calcium salt, absorbing it into the bones where it continues to radiate, leading to irreversible radiation-induced illnesses and a slow death.

of radioactivity after Pierre Curie. The main work of the committee — which comprised Rutherford, Debierne, Curie, Soddy, Hahn, Meyer and Egon Schweidler — involved some hands-on science: they visited Curie's laboratory and then gathered in Lippmann's department in the Sorbonne, where Debierne had set up apparatus able to compare samples of radium by two methods, one of which involved a piezo-electric system developed by Pierre Curie.[55] They confirmed that within the limits of error the two standards to hand — one from Vienna and the other Curie's — were identical.

That was fine as far as it went, but there was a major argument over what quantity of radioactivity should constitute the standard measure. They were in agreement that it should consist of a defined quantity of emanation in equilibrium with radium salt. But just what that quantity might be was another matter. Marie Curie wanted to base it on the output of a gram of radium. She lost the debate on the day, but early the following morning turned up at the hotel where Rutherford and Boltwood were staying and browbeat them over it. Notes in her handwriting, on Hôtel du Grand Miroir letterhead and preserved in Rutherford's papers, laid out the final arrangement: the value would be a gram, and new unit a Curie.[56] Rutherford wrote a short note to Meyer suggesting they 'omit the two sections dealing with the Rad and Curie and report only on the main Radium Standard.'[57] The proposals were passed by the committee, but despite agreeing during discussion that they were honouring Pierre Curie, Rutherford left the attribution of the new unit name ambiguous: it referred to either Pierre or Marie, or perhaps both.[58] Arrangements were made to deposit 20 milligrams of radium with the Bureau International des Poids et Mesures. In this way both the name and the definition of a specific unit of radiation were finally established, but it had been a difficult time and Rutherford came back from the four-day event with a cold.[59]

Rutherford's work at Manchester followed much the same pattern as at McGill and basically picked up where he had left off there. He knew that radioactivity occurred because atoms spontaneously disintegrated, but to explain why that happened he needed to know what comprised atoms and how those components were structured. By this time beta particles were known to be a loose electron. But the identity of the low-energy alpha particle had eluded Rutherford's efforts at McGill, and he now turned his attention to figuring it out. He suspected, but could not prove, that it was a doubly-charged helium ion. Its identity was important because it was clearly a component of an atom, giving clues to its nature, and — more crucially — he intended to use it as a tool to probe atomic structure. Beside this work he also had a fair amount of gap-filling still to do, detailing the way specific radioactive elements broke down and the daughter elements the breakdown produced.

Rutherford structured his research programme carefully: among his papers is a one-page list of 19 topics he laid out for 1908–1909, ranging from specifics such as 'life of Radium D' to broader investigations such as 'scattering of alpha particles'.[60] Rutherford's own initial focus was on alpha particles, for which he wanted an automatic counter. He had tried to develop a mechanism at McGill, without success. Now he was determined to make that happen. One option was measuring particle deflection in a magnetic field, but a magnet of the scale he needed ran outside any reasonable departmental budget. Still, there were other ways to get there. Rutherford's friend Townsend had shown how alpha particles could ionise atoms of a gas by collision. Now Rutherford wondered whether the electrical charge from an alpha particle collision could be amplified in a suitably calibrated magnetic field. He set Geiger on to the job. Geiger's work relied on Rutherford's prior experiments and produced a device that displayed a dot of light on a screen. When an alpha particle entered the apparatus the dot jumped.[61]

This device became a standard mechanism for detecting alpha radiation, and Geiger's name was so inextricably linked to it that although other radiation-detecting devices developed since have used different principles to detect other particles, all are loosely called 'Geiger counters'. Strictly speaking the original, which detected alpha particles, should have been the Rutherford-Geiger counter. But as usual Rutherford was unbothered: to him the counter was a means to an end, not an end in itself, and he had no problem with Geiger getting the credit. Rutherford found that the counter worked extremely well, telling Hahn that the effect was so clear he could show it off to audiences.[62] Later he found ways of recording the count on film.

Rutherford then turned to the nature of alpha particles. He worked with Geiger to repeat an experiment he had done in 1905, this time using one of the radium decay products, 'radium-C' (bismuth-214) as a source of alpha particles. Rutherford intended to make electrical measurements, from which he could identify the mass of the particles. That carried two problems. The first was doing so without skewing the values: he knew that alpha particles could knock electrons out of any residual gas in the equipment, draining the positive charge he expected to measure. That meant he had to find ways of reducing the pressure inside his gear to a hard vacuum — something not easy to achieve. He also had to find a way to get rid of any stray beta particles emitted by the radium. All this was solved by exploiting electrodynamics: he calculated the degree to which the beta particle could be deflected by a magnetic field, making sure the length of the tube was enough to allow this to happen.[63]

The second issue was counting the number of particles involved, for which Rutherford decided he needed a different method than the amplifier counter he and Geiger had developed. Instead he used a phosphorescent screen that would light up when struck by an alpha particle, the same technique he and Thomson had used a decade earlier when investigating X-rays. This decision also meant that any experiment involved hours staring at the screen counting off each dot of light, something Rutherford found difficult. Geiger had more patience. Rutherford was impressed, later telling Bumstead that Geiger had 'worked like a slave'.[64]

The apparatus that emerged was complex — and needed to be, because the results relied on finicky measurements with a Dolezalek electroscope. Rutherford and Geiger even took into account the distortions produced by the small quantity of gamma radiation that Rutherford knew radium produced. The minuscule charges they were trying to measure contrasted with the strength of the electromagnet that lay at the heart of the apparatus. This was adjustable and could draw anything up to 20 amperes, though the experiments were typically run at 12. To put that proportion in twenty-first century terms, even that 12-amp draw was about 1.45 times that of a standard two-kilowatt fan heater at New Zealand's typical domestic line voltages.[65] The fact that Rutherford needed energy of this scale merely to deflect the particle streams produced by a few milligrams of 'radium-C' — which we now know to be bismuth-214 — underscored the forces he was dealing with.

It turned out that the charge on an alpha particle was twice that of a hydrogen atom. On the understanding of the day this meant that the alpha particle was a charged helium atom and became ordinary helium once the charge was lost.[66] Rutherford was able to prove this with the assistance of a new postgraduate student, Thomas Royds, who helped develop apparatus for compressing alpha particles inside a container to the point where it was possible to run a spectral analysis. This principle was well established by that time: it involved comparing the spectrum of light shone through an unknown gas with the spectrum of light that had not. The difference was the 'signature' of the unknown gas. And it turned out that the spectrum of compressed alpha particles whose charge had bled off was indeed the same as that of helium. In short, Rutherford's early guess as to what alpha particles were had been correct.

Rutherford's first year at Manchester was his own annus mirablis. The fact that he had achieved all this in Manchester and had not at McGill is interesting. The larger physics community in Britain had helped, giving Rutherford a range of top-flight graduates and colleagues to work with. As one of his students recalled, his laboratory was filled with 'between fifteen and twenty' students under his 'fairly constant supervision.' He kept close tabs on them with daily rounds of their work and was 'brimful of ideas.'[67] The contamination problem at McGill likely also entered the picture. There is some evidence that the Manchester laboratory had also been contaminated by Schuster's prior work with radium, but this was not in the league of the problem at McGill.

Late in 1908 Rutherford was told he was going to be awarded the Nobel Prize. He had no idea it was coming, though he may have been aware that his friends and colleagues had been trying to get him one. Early in 1907, as a thank you for his work at McGill, John Cox nominated Rutherford on the strength of his work in radioactivity. The nomination came too late to be considered for the 1907 round. What Cox didn't know was that Rutherford had also been nominated five times the same year by German physicists, including Max Planck, while Svante Arrhenius (1859–1927) had done the same in chemistry.[68] In the event the Nobel Committee gave that year's prize in physics to Albert Michelson. Cox tried again in 1908, when Rutherford was also nominated three times in chemistry and five in physics. The message was clear, and the main question facing the Nobel Prize organisation and its subcommittees was whether to award Rutherford one in physics or chemistry, and whether to include Soddy, who was nominated with Rutherford by Oskar Widman.[69] The chemistry committee included Arrhenius, who had essentially founded the field of physical chemistry and championed Rutherford's case. In the end the award went to Rutherford alone, in chemistry, for his 'investigations into the disintegration of the elements, and the chemistry of radioactive substances.'[70]

Rutherford was told confidentially during November, ahead of the mid-December award ceremony. But then the story leaked out of Stockholm, and both Rutherford and Mary found themselves having to evade the attention of the British media. He now felt able to carefully tell trusted friends in Europe. To Hahn he also revealed his surprise at the nature of the award, remarking, 'I am very startled at my metamorphosis into a chemist.'[71]

The leaked news made a splash in New Zealand. A telegram

carrying the story reached the New Zealand Press Association at 8.49 a.m. on 25 November and the news was carried by that day's papers.[72] All kinds of misinformation circulated with it. Rutherford had married a 'Christchurch girl' (true); he had discovered radium (false); he was still at McGill (out of date); and the prize was worth £6,000 (it wasn't).[73] The award was also taken as one for New Zealand: as the *Taranaki Daily News* declared, the Dominion had been 'honoured in the person of Professor Rutherford, who has been awarded the Nobel prize.'[74] Everybody wanted a piece of it. The *Colonist* tied him to his school: 'An Old Nelson Collegian.'[75] Meanwhile the *Press* published a lengthy piece on this 'celebrated graduate of Canterbury College', whose prize had made him part of a 'select company'. The editorial also scoffed at the idea that it crowned Rutherford's career: he was still a young man. Much remained for him to achieve.[76] Even Bickerton was briefly in the limelight, pressed for memories of his student.[77] Only the *Manawatu Times* begged to differ: to them Rutherford was an 'ex-New Zealander.'[78]

Rutherford soon found himself cause célèbre among his colleagues. In theory the leak was ignored: the fact of his prize was officially secret until awarded, but at a Cavendish dinner on 5 December he was feted by Thomson and subject of a song written in his honour by Alfred Robb, to the tune of 'A Jovial Monk Am I.'[79]

When Rutherford and Mary went to Stockholm a few days later for the awards they were taken on a tour of the city, then went to a dinner party hosted by a member of the Academy, Professor Gustav Retzius. The formal ceremony took place on 10 December at the Music Academy, where the prizes were presented by King Gustav V. Rutherford's was introduced by Professor K.B. Hasselberg, president of the Royal Academy of Sciences.[80] Rutherford gave a speech, remarking that of all the transformations he had dealt with in his work the fastest was his own transformation from physicist to chemist.[81] Next day he gave a lecture on the chemical nature of radium. And then he and Mary left for a lightning trip through Europe, calling in to see Hahn and his wife in Berlin, where — as the latest Nobel laureate

— Rutherford was taken around the major laboratories. They went on to Leyden to see Hendrik Lorentz before returning to Manchester where they were reunited with their daughter, Eileen.

More congratulations flooded in. 'My dear Rutherford,' Lord Rayleigh wrote on 22 December, 'A thousand congratulations on your well-earned Nobel prize.'[82] Rutherford also ended up with a cash award that has been variously quoted, but which he told his mother was £7,000,[83] a figure of around £710,000 in mid-2020s values.[84] Either way the figure came to around seven years' worth of professorial salary and put him in a comfortable financial position. Rutherford gave away several hundred pounds to friends and colleagues and stashed the rest. But he had long been a car enthusiast, and the money opened doors. In this first decade of the twentieth century cars were still very much a hobby for the wealthy — expensive to buy, costly to maintain and requiring a good deal of tinkering. The cost would have stretched Rutherford's salary, but his Nobel Prize changed things. In 1910 he bought a four-seater Wolseley. It arrived on Good Friday and, as he explained to his mother, he spent 'three days running around' to practise driving before heading out on a road trip of over 800 km. Rutherford was rather proud of the fact that he had 'learnt to drive fairly well without a single incident, even of running over a chicken.' The car was capable of up to 40 mph (70 km/h) though, as he told his mother, he hoped to avoid speed traps with their ten-guinea fines.[85] Motoring was, indeed, a sport of the rich.

Where next was clear. Rutherford had atomic structure in his sights, the key to understanding why atoms self-transmuted and produced radioactivity. The point gives us further insight into the man. In his first years he had been driven by the need to validate himself. Now he had the highest honours his field could offer. But the mysteries of the atom remained — and both his curiosity and his verve were undiminished. Atoms were going to yield to his efforts, come what may. By this time a number of physicists had been pondering the problem of atomic structure. Thomson's model of 1904 was joined the same year by an alternative concept proposed by the Japanese physicist Hantaro Nagaoka (1865–1950), who devised a 'Saturnian' model in which hundreds of electrons orbited a large positively charged sphere in a flat plane. This proved impractical and Nagaoka discarded it in 1908, but the question of just what constituted an atom was still dominating the wider physics community.

Rutherford knew that electrons were part of the story. The question was how all these components and electric charges fitted together: any answer was going to have to be consistent with empirical data. Experiments had already shown that beta particles — which Rutherford knew were electrons — could be scattered by firing them through thin metal foils. There was no scattering data for alpha particles, but he now knew they were positively-charged helium atoms. In terms of what was known at the time, this implied they were also the size of atoms, so he decided to collide alpha particles with various foils and see what happened. The technique involved firing a thin stream of alpha particles through the foils, then observing the scattering pattern on a small phosphorescent screen via an eyepiece. The apparatus could be adjusted to observe what happened from various angles. The basic principle — smashing particles into each other and getting data from the way they scattered — became the basic tool by which physicists explored the subatomic world, and is still used today by the Large Hadron Collider operated near Geneva by CERN, among other similar instruments around the world. All draw their origins, in concept, from Rutherford's laboratory equipment.

The only problem from Rutherford's point of view was that the observations took hours, demanding more patience than he had himself. He put Geiger and a young undergraduate student, Ernest Marsden (1889–1970), on to the job under his close direction. His instructions included, in the end, a request to see whether anything bounced straight back. Early in 1909, either Geiger or Marsden — depending on source — reported to Rutherford that they had discovered exactly that. Rutherford knew the scattering was caused by the Coulomb force — electrodynamic repulsion of like charges. But the discovery that something as massive as an alpha particle

EVOLVING ATOMIC MODELS 1808–1911

1. Classical physics: Dalton 1808

Dalton's atomic model was based on classical physics: to him atoms are solid, indivisible, and atoms of same elements are identical.

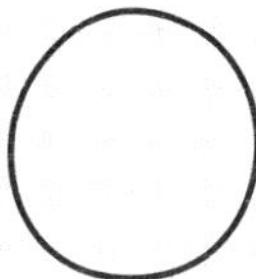

2. Electrodynamics introduced: Thomson 1904

Thomson's 'plum pudding' model introduced electrodynamics and the first subatomic particle: he suggested atoms consist of a spherical positive charge within which electrons circle in defined patterns.

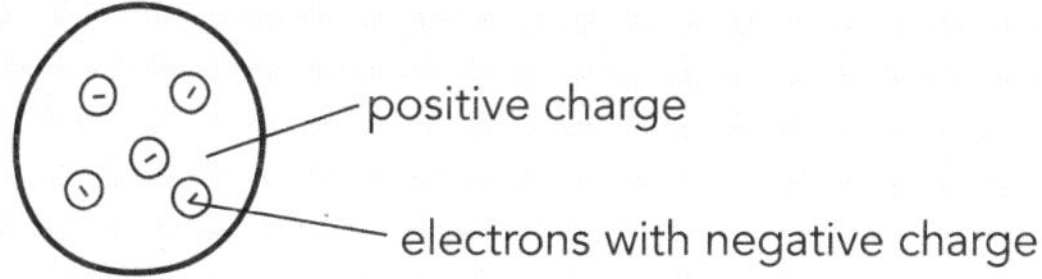

3. Alternative electrodynamic ideas: Nagaoka 1904

Nagaoka's 'Saturnian' model suggested atoms consisted of a spherical positive charge around which electrons orbit in the fashion of Saturn's rings. Nagaoka swiftly decided this was incorrect.

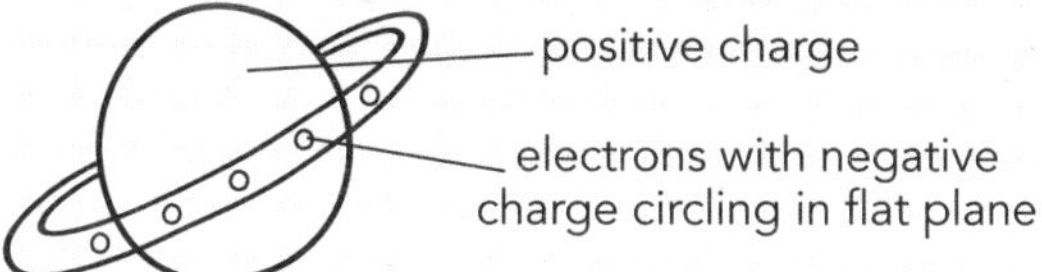

4. Nuclear breakthrough: Rutherford 1911

Rutherford's nuclear model proposed that atoms are primarily empty space with most of the mass concentrated in a tiny positively charged nucleus. This 'nuclear' concept remains in place today. He did not specify electron orbits in his initial publications, knowing this required more work.

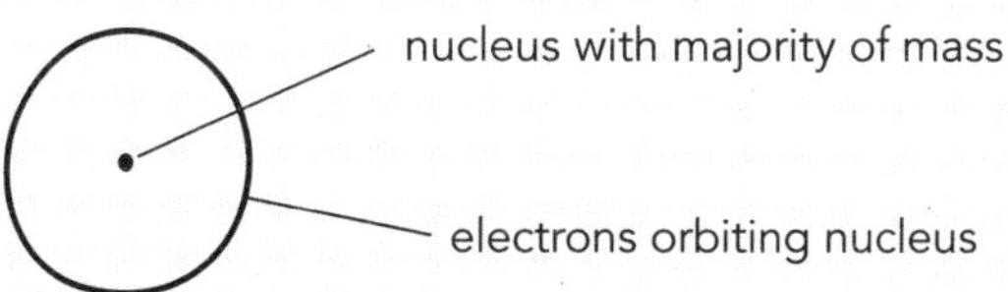

could apparently ricochet straight back the way it came posed more questions than it answered. Just what this meant was not clear just then, but he knew that the energy required to bend the vector of an alpha particle around even 90° was enormous, relatively speaking. Geiger and Marsden wrote up the results and sent them to the Royal Society in May, but their paper was purely informational.[86]

Much of Rutherford's time during 1909 was taken up on other matters, including work with Boltwood, who had come across from the United States for a period. As time went on he brought others on board, including Charles G. Darwin — grandson of Charles — as his mathematician, the Hungarian physicist George de Hevesey, and the brilliant young physicist Henry Moseley. The whole physics community was buzzing with excitement, swapping data with each other as they pushed forwards into the unknown. Sometimes results were as human as they were scientific. 'With regard to the emanation which you so kindly sent me,' Hugh Callendar wrote once to Rutherford, 'I have stupidly left my observations at the Garage & don't like to fetch them in this thunderstorm.'[87]

All this time the problem of atomic structure bubbled along in Rutherford's mind. He knew that alpha particles were charged helium atoms, and he knew helium gas didn't have the same effect. What did that mean? There was also the nature of the dispersal. Only a few particles were bouncing back but many were being scattered. He was not the only one looking into the issue. Thomson had been trying to explain scattering in terms of his own 'plum pudding' atomic model for some time. At this point the sole known subatomic particle was the electron, and Thomson focused on reconciling that with both his 'soup' model and the new experimental observations. Thomson's model initially produced an assumption that atoms might include hundreds of thousands of electrons.

Thomson steadily found ways of whittling the number down, and in 1910 the physicist James A. Crowther (1883–1950) ran experiments that seemed to verify that work, leading to the idea that atoms had around three times the number of electrons as their atomic weight.[88] The expected structure remained the same, a solid blob of electromagnetic charge within which electrons circled. However, that carried a major weakness. The act of moving in a circle meant the electrons also bled energy, because they were constantly changing direction — technically, their 'velocity vector'. This required energy, an energy that could come only from the momentum of the electrons themselves. In short, Thomson's model was unstable, and he struggled to explain ways around it.[89]

Rutherford wasn't convinced. He believed the alpha particle was physically the size of an atom. We know now that this wasn't quite true, but it was a reasonable assumption at the time; and, logically, the alpha particles fired in the scattering experiments should be shoving atoms aside like a cue ball. A few, naturally, would bounce straight back. But the issue worried him. He checked with colleagues. Nagaoka was visiting England and came to see Rutherford in Manchester, where the Japanese physicist was 'struck by the simpleness of the apparatus you employ and the brilliant results you obtain.'[90] Late in 1910 Rutherford decided to approach the problem in two ways, first from the perspective of the electrodynamic energies required to deflect alpha and beta particles, and statistically in terms of the known number of atoms in any given volume of material. He began scribbling calculations in his notebook, and it did not take long to realise what was happening. The key insight apparently arrived without conscious thought one Sunday afternoon in early 1911, about half an hour before Charles G. Darwin and other colleagues were due to arrive for supper.[91]

Later, Darwin remarked that he did not think Rutherford waited for the meal before telling everybody; and we can imagine the scene in which the excited New Zealander — whose voice was loud at the best of times — boomed out the news. The solution revealed his mastery of both theory and mathematics. The trajectory of the scattered particles was hyperbolic, a shape Rutherford now remembered was defined by the geometry of a cone. This meant the scattering could be analysed by well-established conic equations. The alpha particle was far too large to be deflected by an electron. Therefore atoms must contain something fairly massive and yet also, clearly, very small because not all the alpha particles were being flung straight back.

That idea was radical because it meant the atom could not be solid. This also implied there was at least one further type of particle inside it. Rutherford realised that a good deal of data could now be extracted from the trajectories of the alpha particles. Once again — just as Cook's measurement of Venus' eclipse unlocked a raft of data about the solar system — a single piece of information, in this case the connection to conics, could unlock a whole lot of unknowns about the atom. Darwin, writing 27 years later, recalled that Rutherford asked him to cross-check the mathematics when he got to work the next day.[92]

Rutherford's concept of a largely empty atom with minuscule nucleus that carried most of the electric charge and mass was a

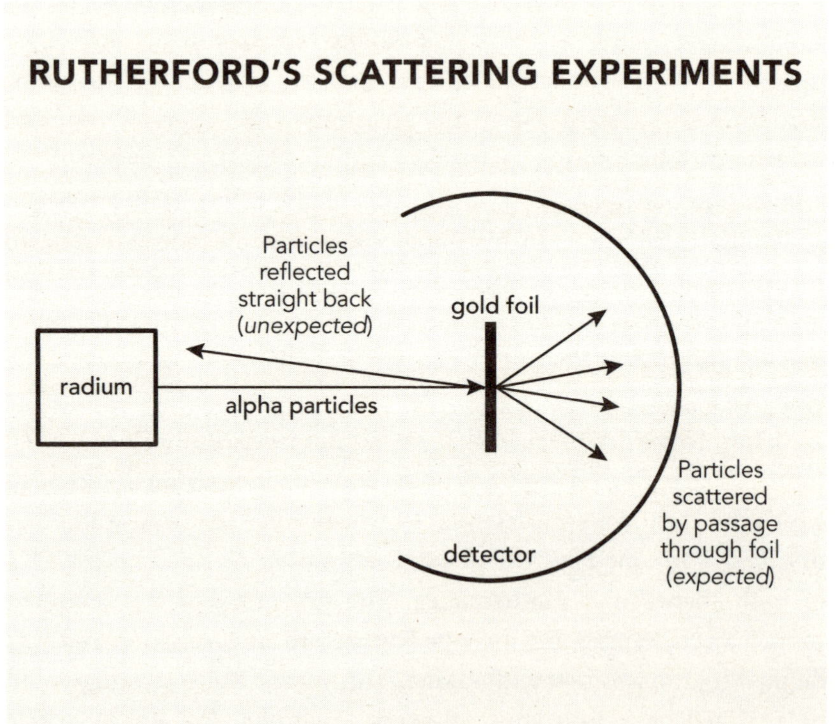

fundamental shift of concept, and one that has never been dislodged. The specifics of how the electrons orbited have been revised, but the basic idea of a nuclear structure holding the bulk of the mass has not. It was a radical thought by 1911 standards. He didn't yet have empirical detail as to where the electrons were, other than orbiting in some fashion; but the discovery of the nucleus and its scale alone was dramatic enough. The issue of again being pipped at the post worried him, so he prepared an interim paper for the *Philosophical Magazine*, which was published in April 1911. He was explicit: this was preliminary and his work was ongoing.[93] And he carefully referred only to what had been defined to date. The implication was that there was another subatomic particle besides the electron, but Rutherford described his findings purely in terms of electrical charge.[94] He had more, yet deliberately omitted the details until he had further empirical proof.[95]

This caution was very typical of Rutherford. His new idea was radical, and much remained to be determined when the paper appeared. One initial question was whether the central charge was positive or negative. Rutherford discussed that with Bragg by correspondence, then called on Geiger and Marsden to conduct experiments. This included determining the specific angles at which the alpha particles were reflected back, from which the diameter of the central charge could now be calculated thanks to his Sunday afternoon insight. The extent of Rutherford's direct influence on this work is clear: Geiger even accepted all Rutherford's suggestions on the final paper.[96] Rutherford's own workings, headed up 'Theory of structure of atom', amount to more than 50 pages of equations and rough notes, undated but appearing to be preliminary material ahead of his formal papers on the topic, including calculations for the dimension of a nucleus.[97]

These calculations revealed just how radical Rutherford's concept was. When he analysed the scattering curves he found that the central charge occupied a space 10,000 times smaller than the atom itself. In other words, atoms — the atoms that classical physics had always portrayed as solid — were basically empty space. This also explained how alpha particles worked as a research tool: they were a bare helium nucleus and thus far smaller than a helium atom. Of course this posed fresh questions. One was the nature of electron orbits. Another was how solid matter could exist if its components were mostly empty space. Neither could be fully answered in 1911 — it took the development of a whole new branch of physics. Eventually, in 1925, armed with what Rutherford discovered about atomic structure, the physicist Wolfgang Pauli showed that solid matter exists because two electrons cannot occupy the same quantum state. Atoms therefore cannot interpenetrate. But that insight relied on Rutherford's nuclear structure, showing just how fundamental his 1911 discovery of the atomic nucleus was to the whole of physics.

This was not the only direction Rutherford took during this period. Even as he delved into atomic structure he launched another left field approach to geological time spans, this time determining the age of spall marks in minerals, damage known to be caused by alpha particles. This was typical of Rutherford: he never stopped asking questions — and a question posed was a question that had to be answered. Amidst all of that he was also wrapped in an effort to support Charles Phillips' research into radiotherapy at the Cancer Hospital in London, where the standard radium measure Rutherford had assembled became a yardstick.[98] All this came at a cost to both his hands-on work and the guidance he was giving to his team. Letter-writing became a chore and he finally hired a stenographer, much to the relief of his correspondents. 'It adds a new pleasure to the receipt of your letters,' Boltwood declared. 'That of being able to read them on the first trial.'[99]

Rutherford's work may have seemed scattergun, but in the first years of the twentieth century the unknowns far exceeded the knowns, and Rutherford had the uncanny knack of being able to seize on something others saw as unimportant, showing how it answered profound questions. His discovery of the atomic nucleus from the way some particles occasionally reversed direction was but one example.

To Rutherford, in short, all information was valid, and there was no way of knowing how one line of enquiry might interlink with another until all had been explored.

As always it took time for the physics community to accept Rutherford's ideas about an atomic nucleus. It was understandable: he had only partial supporting data and many questions remained unanswered. Nonetheless, his bold proposals for atomic structure added to the many ideas floating around the physics community.

By this time the new physics was in flux. Maxwell's electrodynamics were couched purely in classical principles, but the effort to reconcile what was being observed with those ideas was stuttering. There was no question that electrodynamics applied, but what was being observed increasingly couldn't be explained in terms of older classical paradigms. Planck's radical idea of energy steps — quanta — at the scale of particles answered some of the issues, but had gained no traction. Even Planck was dubious. Einstein's 1905 paper applying the same principles to the photoelectric effect was initially over: Einstein remained a minor figure. But the questions and inconsistencies continued to pile up.

During 1909–1910 the German physicist Walther Nernst (1864–1941) began looking into the idea of energy quanta as a way of proving his 'heat theorem'. In the process he discovered Einstein's 1905 paper on the photoelectric effect. Nernst felt that Planck and Einstein were on to something. He went to see Einstein early in 1910 and was impressed. Shortly afterwards he decided that an international conference should be held to discuss the new ideas.[100] This was easier said than done. Nernst was reluctant to be seen as originator and had no way of funding it. Then in July 1910 he approached the Belgian chemist Ernest Solvay (1838–1922), who in the early 1860s had devised a process for producing soda ash (sodium carbonate), used in a number of large-scale industrial processes, including glass manufacture. By 1900 around 95 percent of the world's supply was made with his method.[101] Solvay had buckets of money as a result and, it turned out, buckets of enthusiasm for science. The conference took nearly a year to organise, but by May 1911 Nernst had shaped up a provisional list of participants.[102]

In the end 18 physicists, including Rutherford, Einstein, Planck and Lorentz, arrived at the Grand Hotel Métropole in Brussels on Monday 30 October for a five-day discussion. The conference took the grand title 'The theory of radiation and quanta'. Sessions were chaired by Lorentz and were built around papers submitted by several participants. The focus was on quantum matters and their relationships with what was being observed: Rutherford's proposals about atomic structure were never mentioned.[103] Rutherford did not submit a paper but joined discussions on one by the German physicist Arnold Sommerfeld (1868–1951) that discussed energy quanta, illustrating his points with photographs from scattering experiments.[104] Louis de Broglie, who had the job of assembling all the material for publication, later sent Rutherford drafts of the conference report for comment.[105]

The conference had a major effect on the shape of modern physics, making the idea of energy steps (quanta) more respectable and opening the door for ongoing investigations. Some participants became converts, among them Maurice de Broglie and Henri Poincaré. The discussions, though brisk and somewhat heated at times, had been so successful that Solvay decided to fund an organisation to host regular conferences. He provided one million francs — about four million Euros in 2024 values[106] — and the Institut International de Physique Solvay was formally established from 1 May 1912. Rutherford was on its scientific committee.[107] The conferences this organisation subsequently organised had a profound effect on the development of modern physics.[108]

The 1911 conference also summed up the cultural reality of physics at that time. Despite the fact that the Western scientific approach had been embraced by top class mathematicians and physicists in places such as India and Japan, the dominant figures remained those in Western Europe and its current and former colonies, which included the United States. Representation at the first Solvay conference was further skewed even within that diaspora: just two physicists from England attended and there was nobody from Canada or the United States. The community was also primarily male. The only woman was Marie Curie, and it was a sign of the times that the media were more interested in the fact that she was in a relationship with fellow physicist Paul Langevin.

Rutherford returned to Britain with questions about the deceleration of particles when passing through matter.[109] The conference was still top of his mind a few weeks later when he was invited to an annual dinner at the Cavendish, where he met one of Thomson's students, a Danish prodigy by the name of Niels Bohr who was there on a Carlsberg scholarship. They got talking. Rutherford gave what Bohr later recalled was a 'vivid account' of the Solvay conference.[110] The upshot was that Bohr decided to transfer to Manchester, where Rutherford allowed him to focus on theory. It was an inspired decision: Bohr's strength was not in hands-on work and Rutherford needed a theoretician. The problem Rutherford had with his atomic model was the same as Thomson's: by classical principles, electrons bled energy when circling a nucleus, meaning they would quickly collapse into the centre. Obviously they weren't doing that, which to Rutherford meant more data was needed. He put Bohr on to the problem of particle deceleration when encountering matter and, meanwhile, asked Darwin to look into scattering and electron orbits.

Bohr then found the answer to the electron problem by picking up the idea of energy steps at atomic level, which he saw as a way to solve the instability of Rutherford's atomic model. Bohr remained at Manchester until July 1912, when he left for Denmark to get married. He subsequently took up a position at the University of Copenhagen, but wrote regularly to Rutherford, leaning on him for comment and input into a series of papers — to Bohr, 'chapters' — that resolved the problem of electron stability.[111] At one point Bohr asked Rutherford to run some experiments that hadn't worked in Copenhagen.[112] Then, at the end of March, Bohr and his wife visited Rutherford in Manchester, where he was able to discuss the questions directly.[113]

Arthur Eve later described meeting Bohr at Rutherford's house: to Eve the Danish physicist presented as a 'slight looking boy.'[114]

The first of three papers describing Bohr's quantised atom — published in the *Philosophical Magazine* in July 1913 — explicitly thanked Rutherford for his 'kind and encouraging interest.'[115] Bohr's idea was simple in concept: he took Planck's idea of energy quanta and ran with it. Electrons orbited Rutherford's central nucleus, Bohr argued, in paths defined by their energy, stepping from one to another. When an electron stepped from a higher energy level to a lower one it emitted that energy at a frequency defined by Planck's equations. This explained the light and X-ray spectra that were being observed in what Rutherford called 'excited' atoms.[116] The specifics, as always, were harder to pin down — not least because Planck's constant, a value that lay at the heart of Planck's ideas, wasn't fully defined. Bohr initially also mixed classical and quantum principles which, as Rutherford gently pointed out, 'make it very difficult to form a physical idea of what is the basis of it ... how does an electron know what frequency to vibrate at when it passes from one stationary state to another?'[117]

Bohr took on board both that and Rutherford's plea to slash the length of the paper by at least a third. Bohr's final concept did not address every issue, but the model was stable with its quantised electron orbits. It later became known as the 'Rutherford-Bohr' model and was seen as a plausible explanation demanding further study. Bohr spent considerable time afterwards looking for ways to validate his proposals against what was being found.[118] Rutherford certainly accepted that Bohr had found a way forward. But he knew that much remained to be discovered, including the nature of the nucleus.

This discovery went hand-in-hand with a new field of investigation: crystallography, the art of firing X-rays at crystals and observing the way the rays were diffracted. This was conceptually identical to the way Newton used prisms to examine light, but at far higher frequencies, offering a new way of investigating atomic structure. At the time it was radical. Years later, in 1935, Rutherford made the point

RUTHERFORD'S NUCLEAR ATOM WITH BOHR'S ELECTRON STRUCTURE

RUTHERFORD'S STUDENT Niels Bohr introduced early concepts of quantum mechanics to show that electrons stably orbited Rutherford's nucleus in discrete layers. While modern diagrams show the nucleus components as protons and neutrons, neither were known when Bohr developed this model. However, the mass and charge on the nucleus were available.

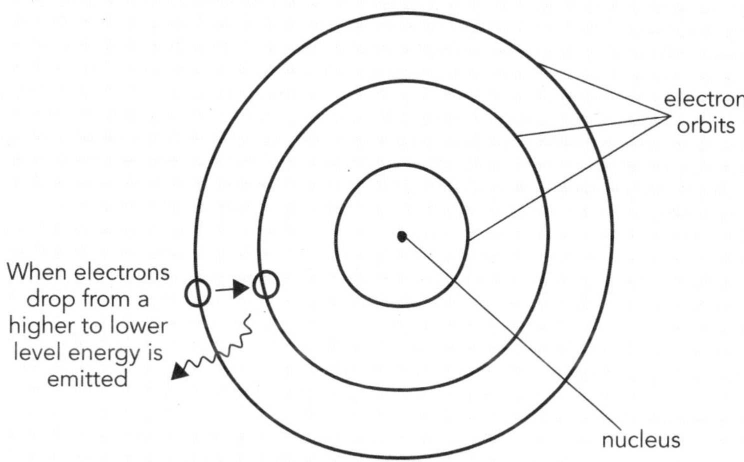

clear: 'We can hardly realise today how strange and revolutionary this new theory of atomic spectra appeared to many physicists at the time … a complete break with the ordinary ideas of classical physics.'[119]

These matters became the main theme of the second Solvay Conference, held in Brussels at the end of October 1913. Thirty physicists attended, again including Einstein. Once again representation was skewed towards Western Europe, though there was a larger British contingent that included Rutherford, Thomson, Bragg and Jeans. This time the focus was on the structure of atoms and on crystalline structures. Thomson produced an extensive paper on atomic structure, still referring to 'corpuscles' and provoking discussion on the relationship between particle scattering and the conclusions about structure.

Rutherford presented his own concept of largely empty atoms with a tiny but relatively massive nucleus during this debate. In 1911 he had cautiously referred to a central charge, but now confidently referred to '*un noyau central de dimensions excessivement petites, portant une charge positive et contenant la plus grande partie de la masse de l'atome*' ('a central nucleus of exceedingly small dimensions, carrying a positive charge and containing the greater part of the mass of the atom').[120] The only subatomic particle then known was the electron, and debate turned to whether these might be present in the nucleus and how they might account for radioactivity.[121] It was a brisk argument.[122] However, we know only what the stenographer recorded, verbatim material then filtered through editorial work and translated into French. There were also other discussions: they all stayed in the same hotel and there was a good deal of informal chatting.[123]

Rutherford came away from the conference with a new focus on crystallography and approached the Chemical Laboratory at Cambridge for suitable material. William Pope, then Chair of Chemistry, gave him some samples of rock salt and offered topaz or quartz if needed, adding: 'There is a good diamond of the size you want amongst the Crown Jewels — the Kohinor.'[124]

In early 1914 Rutherford put together a paper detailing material he had 'purposely omitted' from his 1911 publication, and adding new research completed since by Geiger and Marsden that was consistent with the idea of a nucleus and thus 'indicated the essential correctness of this hypothesis.'[125] Bohr's quantum proposals were 'of great interest and importance to all physicists' because they were the 'first definite attempt to construct simple atoms and molecules and to explain their spectra.'[126] Much remained to be discovered about the nucleus, but he could now more confidently describe the position of electrons, which he visualised orbiting the nucleus like planets, 'extending to distances from the nucleus comparable with the ordinary accepted radius of the atom.'[127]

At the time Rutherford's nucleus with Bohr's quantised electron orbits remained controversial — and he knew it. Part of the problem was that available tools were limited. More crucially, explanations were couched in terms of electrodynamics until proven otherwise. This was valid enough. But as Bohr had shown, quantum mechanics also applied — and that field was still in its infancy. What physicists of the day didn't know, though Rutherford clearly suspected, was that other forces were also at play, and electrodynamic answers would always be incomplete. The discovery that electromagnetism operated according to quantum-mechanical processes, not classical mechanics, went some way towards answering the questions. However, the two further forces operating at atomic and subatomic level — the strong and weak nuclear forces — were not discovered until decades later.

Given all these unknowns it is all the more remarkable that Rutherford, using incomplete data and purely classical electrodynamics, correctly showed that atoms were mostly empty and that they contained a nucleus holding most of the mass and charge. The details gained dimension with quantum mechanics, but the concept of nuclear structure was Rutherford's. The concept of atoms as mostly empty space with electrons whirling around a tiny nucleus reversed Dalton's ideas, threw a wrench through every atomic model proposed so far under electrodynamics, and established a new basis for explaining how chemistry worked. But for Rutherford it remained only a beginning. The nuclear structure he had identified in atoms was an important step in his primary goal of understanding radioactivity. But it was only a step, and many questions remained. The mysteries of the nucleus beckoned.

9 | Disintegrating atoms

The rush of your advance is overpowering, and I do not wonder that Nature has retreated from trench to trench, and from height to height, until she is now capitulating in her innermost citadel.

— George E. Hale to Rutherford, 1 June 1914[1]

ERNEST RUTHERFORD WAS ON TOP of the world by the end of 1913. A golden year was capped by a knighthood in Britain's 1914 New Year's honours list. He told de Hevesy that it was 'very satisfactory to have one's work recognised by the powers that be,' but felt a knighthood was 'a little embarrassing for a relatively youthful and impecunious Professor.'[2] His parents thought otherwise, telegraphing: 'Loving congratulations.'[3]

The news swiftly made the papers in New Zealand.[4] Once again it was a shared glory — as the *Otago Witness* declared, a 'tribute to the quality of New Zealand citizenship.'[5] The *Southland Times* called Rutherford 'New Zealand's most brilliant son,' who had 'won the Knighthood ... by sheer merit.'[6] New Zealand would, the *Evening Star* announced, be 'particularly gratified with the knighthood conferred on Dr Ernest Rutherford, a native of Nelson.'[7] The *Colonist* insisted that New Zealand was 'proud of Professor Rutherford, whose valuable contributions to science, and especially his research in radio-activity, doubly merit the guinea stamp — this last a reference to a Robbie Burns poem exalting the idea of personal integrity. His career had been 'followed with interest and satisfaction by all who rejoice to see New Zealanders distinguishing themselves abroad.'[8] Only a few commentaries focused on the man himself. Rutherford, the *Fielding Star* declared, had 'fully earned the knighthood.'[9]

Rutherford spent the first weeks of 1914 completing his second paper on atomic structure, published in March. His next focus was the nucleus. By this time other pieces of the jigsaw puzzle were emerging elsewhere. Soddy, now at the University of Glasgow, was able to resolve a problem with the number of radioactive elements. The periodic table was more than just a diagram: it was a tool that showed how the elements inter-related. There were 11 slots for radioactive elements, but by 1913 around 40 had been identified via decay chains. Soddy was able to show that many of them were actually versions of the same basic element.

More insights into the structural relationships of the elements came from Rutherford's team, notably Henry Moseley — Harry to his friends. Moseley was a talented Oxford graduate who came to Manchester as a physics demonstrator in 1910. He became a graduate research assistant under Rutherford, and in 1913 began working on X-ray crystallography, a relatively new field pioneered by Bragg. Moseley fired X-rays into crystals of potassium ferrocyanide 'mounted on the prism-table of a spectrometer' and watched the way they scattered.[10] Rutherford kept close tabs on what was happening: Moseley reported back to him regularly.[11] The results were decisive. Up to this time chemists had arranged the periodic table broadly by chemical property. Moseley, working with Darwin, discovered a direct relationship between atomic number and the wavelengths of X-rays. This was the first time the periodic table had been found to

ATOMIC WEIGHT, RADIOACTIVE DECAY CHAINS AND ISOTOPES

RUTHERFORD'S PIONEERING work on radioactive decay chains, alongside work by other physicists exploring the same field, produced a lengthy list of decay products that were initially given the names of the element that had produced them, followed by a letter — 'radium A', 'radium B' and so on. The same was true for the other radioactive elements, producing an enormous number of 'radioactive elements', more than the periodic table predicted. Subsequent work, notably by Soddy but also conducted independently by Kazimierz Fajans, showed that most of these were actually variants of a much smaller range of elements that had differing atomic weights, but which were chemically identical. Soddy dug into the question in 1910–13 and coined the word 'isotope' to describe them. Soddy's discovery was all the more remarkable because, as we know today, the full explanation of the principles behind isotopes involves the particles that comprise the nucleus — none of which were known when Soddy made his main discoveries on the basis of laborious chemical analysis alone.

Rutherford's 1911 discovery that atoms have a nuclear structure, followed by his 1917–18 identification of the proton and subsequent prediction of the neutron, joined work by Rutherford's student Henry Moseley and others to both explain how isotopes are formed, and to create the current system by which these isotopes are named.

Rutherford's later prediction of the neutron, which he got James Chadwick to investigate, provided the final piece of the puzzle. Today we know that each element is defined by the number of protons in its nucleus. With the exception of protium — the most common hydrogen isotope — all elements contain varying numbers of neutrons. The number of neutrons defines the isotope of the element: for example, deuterium, an isotope of hydrogen, has one neutron. Tritium, another isotope, has two. Because protons and neutrons have similar mass, this alters the atomic weight of the element.

TABLE 1 Isotopes of hydrogen*

Nuclide	Common name	Protons	Neutrons	Mass (rounded to 3 places)
H1	Protium	1	0	1.008
H2	Deuterium	1	1	2.014
H3	Tritium**	1	2	3.016

* First three isotopes shown. Seven exist in all, but those above tritium have half-lives measured in yoctoseconds, i.e. 10^{-24} seconds, and technology to detect them did not exist until the mid-to-late twentieth century.

** Unstable, half-life 12.33 years, decays by beta emission to helium-3.

TABLE 2 Historic versus modern decay chain names

Historic name	Modern name	Half-life (approx.)
Radium	radium-226	1600 years
Emanation (Radon)	radon-222	3.82 days
Radium A	polonium-218	3.09 minutes
--	actinium-218	2.5 seconds
--	radon-218	35 milliseconds
Radium B	lead-214	26.8 minutes
Radium C	bismuth-214	19.9 minutes
Radium C	polonium-214	164.3 microseconds
Radium C	thallium-210	1.3 minutes
Radium D	lead-210	22.2 years
Radium E	bismuth-210	5.01 days
Radium F	polonium-210	138.37 days
--	mercury-206	8.32 minutes
--	thallium-206	4.2 minutes
Radium G	lead-206	--

have a basis in physics. Moseley's work also showed there were gaps in it, confirming suspicions in the chemical community that there were elements yet to be discovered. And to cap it off, his discoveries endorsed the Rutherford-Bohr atomic model. It was an extraordinary step, providing yet further evidence for Rutherford's concept of the nucleus and showing that Moseley was the brightest and most promising young physicist of his day.

Rutherford had little time for hands-on work himself that year: he was too much in demand. But this gave him the opportunity to reveal where his thoughts were going. He went to the United States in April to present the first William Ellery Hale lectures for the National Academy of Sciences. The short tour took him to McGill, Princeton, Yale and Columbia. This series had been initiated by the astronomer George Ellery Hale and his two siblings in memory of their father. Rutherford was asked to present the inaugural lecture, 'The constitution of matter and the evolution of the elements.'[12] Rutherford used the event to preface where his ideas were taking him. The fundamental property of an atom, he declared, was not its atomic weight but its 'nuclear charge.'[13] Were elements built from some fundamental substance, perhaps hydrogen, as the chemist William Prout (1785–1850) had suggested? Possibly, he declared, speculating that atoms might be composed of 'hydrogen nuclei' and 'negative electrons.'[14] This idea came from the work he had directed Geiger and Marsden to undertake and was remarkably prescient. Rutherford eventually discovered that the hydrogen nucleus was indeed a component of the nucleus in all other elements he found, which he dubbed a particle 'proton' in reference to Prout. In 1914 there was still no proof, and Rutherford was going to have to get it — but once again he had a clear picture in his mind.

Hale was overjoyed, telling Rutherford afterwards that he had 'stirred the imagination of every investigator ... stimulated them to try and do likewise ... I have wondered a hundred times since how I could learn to think and do things in such a big and effective way.'[15]

At the same time Rutherford also prepared for a major trip to Australia and the 1914 meeting of the British Association for the

Advancement of Science. This was a big deal, underwritten by Australia's state governments, with 300 scientists invited from elsewhere and a large contingent of Australian scientists expected to attend. Arrangements had been in train for three years and — thanks in part to a request from the New Zealand government — Rutherford was an organiser at the British end.[16] He was also participating and had to juggle all this with his other work, including arrangements he was making with Meyer for a European physics congress in early 1915. He had time on 29 June to write a lengthy response to Meyer's proposals, reminding his friend that he was about to leave for Australia and New Zealand — and adding that he was 'very distressed by the tragic news of the Archduke and his wife,' which he had just read in the morning paper.[17]

Rutherford, Mary and their daughter Eileen set out for Australia on 1 July, travelling first class aboard the brand new Aberdeen Line steamer S.S. *Euripides*.[18] They reached Cape Town on 20 July.[19] The journey kept them out of a growing crisis in Europe, triggered by the Sarajevo assassination but driven by tensions that had been building for years. Rutherford and his family were in the Indian Ocean at the beginning of August when news came that war had broken out between France, Russia and the central powers. Britain's alliance terms with France did not obligate them to join — but by the time the steamer reached Albany the British, too, had committed to the conflict.

The war was not expected to last long, and for a while the intense series of tours, lectures and discussions in Australia allowed the scientists to focus on their work. Rutherford presented his theory of atomic structure and — informally — remarked that instead of splitting the atom as the next major goal, physicists should look at splitting electrons.[20] But the war soon intruded. A subsidiary gathering in New Zealand was cancelled. Mary's younger brother was an officer in the Territorial forces and was called up for an expeditionary force, so Mary hastened to New Zealand with Eileen to see him before he left for camp. This was no light decision. The powerful German East Asiatic Squadron, under Rear-Admiral Maximillian Graf von Spee, was thought to be at large in the Pacific. New Zealand's war plans were built around the likelihood of German cruiser raids into local waters, fuelling expectations that von Spee would turn up, adding a spice of danger to trans-Tasman travel.[21]

Rutherford joined his family a little later, arriving in Auckland where he was interviewed by the media and asked about atomic weapons of the kind described a few months earlier by H.G. Wells in his novel *The World Set Free*. Wells imagined atomic-powered aircraft carrying 'atomic bombs' that 'would continue to explode indefinitely,' burning craters into the ground and irradiating wide areas.[22] This vision of amped-up radioactivity was not too far from the 'dirty bomb' concepts of the later Cold War. But in 1914 Rutherford hastened to quash speculation. Nobody yet knew how to alter the rate of radioactive transformation the way Wells imagined.[23]

Once again everybody wanted to bask in Rutherford's glory, and he was busy during the early weeks of his New Zealand visit, lecturing, running demonstrations and talking with university officials. He was very much public property: at a civic reception in Christchurch on 13 October the mayor, Henry Holland, told assembled crowds that Rutherford was 'probably the most distinguished scientist in the world today,' and it was 'especially pleasing to remember that Sir Ernest Rutherford was a New Zealander and that Lady Rutherford came from Christchurch.'[24] Finally there was time for family. Rutherford's parents were retired, still living in Taranaki. Rutherford and Mary also spent time in Christchurch, where — as he told Boltwood — they had a 'very pleasant holiday.' They sent Eileen to school 'to occupy her time' and otherwise led a 'lazy life.'[25]

They booked passage back to Britain on the liner *Niagara*, then on the Sydney–Vancouver run via New Zealand. Von Spee's whereabouts were still unknown, which worried Rutherford. He told Boltwood that *Niagara* had the speed to outrun von Spee's flagship *Scharnhorst* 'if she gets on our trail as the latter's bottom must be pretty foul by this time.'[26] As it happened, just two days after Rutherford wrote these words von Spee turned up off the coast of Chile, where his force defeated a British squadron near Coronel.[27] It took some days for that news to spread, but it meant von Spee was known to be too far off to threaten the New Zealand–US route when Rutherford and family left New Zealand on 1 December. The risk now became the Atlantic passage and particularly the approach to British waters. This area was home to German submarines, though at this stage in the war U-boat commanders were expected to surface, request a ship to stop, and board in search for war materiel. In the event Rutherford and his family reached England without incident in January 1915.

By this time the war in Europe had stalled, the armies deadlocked on a front stretching from Switzerland to the Channel coast. Britain's pre-war force had been largely expended in the fighting of 1914 and a new force of citizen-soldiers was being raised, including a fair number of Rutherford's colleagues and his laboratory support staff. He complained to Boltwood that he found it 'difficult to get people to work very hard ... as most of them are engaged in munitions or other work.'[28]

War was the last thing Rutherford wanted. To him science knew no nation. Now, as the German chemical industry repurposed its dye-making capacity to produce poison gas, and as British scientists looked to similar technologies, the world seemed to have gone mad. Good friends and former colleagues from the central powers, including Geiger and Hahn — whom he treated as family, and who viewed him as a father figure, friend and mentor — were legally now the 'enemy'. None of them liked the idea, and Rutherford tried to keep in touch, typically using the American consulate as an intermediary. Meyer responded to his first approach with a chatty reply and they kept up an intermittent correspondence, which retained the tone of their pre-war frankness.[29] Geiger responded with a letter that read more like that of a son writing home to a parent than that of a serviceman communicating with someone now deemed the enemy. He even explained how he had recently been at the front.[30]

Rutherford's colleagues were also dispersed. Some of his English students ended up in German hands, among them James Chadwick, who was working with Geiger in Berlin when war broke out and was interned. Rutherford was able to make contact with him, as did Geiger.[31] Chadwick was eventually able to get a place in the camp where he could pursue scientific work.[32] Andrade ended up serving in the artillery in France.[33] Darwin also ended up in France, where he worked with Lawrence Bragg, the physicist son of William, on a method for sound-ranging on enemy artillery.[34] They kept in regular contact through the war.[35] Marsden, meanwhile, went to New Zealand to take up professorship at Victoria University of Wellington, arriving there in mid-March 1915.[36] Rutherford kept in contact, sending him money to support his work.[37]

One of Rutherford's early worries was that somebody might find a way of weaponising the enormous energies trapped within atoms: he knew Wells' vision of pepped-up radioactive output was impossible on what he and his colleagues currently knew, but was also well aware that they had much still to discover. Thanks to back-channel contacts he knew his German friends weren't actively looking into the idea. Still, the prospect that somebody might find a way to release atomic energies for war could not be ignored.[38]

The middle months of 1915 brought multiple blows. The first came in August when Moseley fell to a sniper bullet at Gallipoli. He was just 27.[39] Rutherford had last heard from him in April, when Moseley wrote to say that he was in 'splendid health' working in signals, and that his failure to publish his undergraduate university paper 'lies heavy on my conscience.'[40] Now he was dead, and Rutherford told a colleague that the news 'plagued me with grief.'[41] Moseley had been 'one of the best research students that I ever had.'[42] Others thought the same: Andrade told Rutherford he was 'most distressed ... The only consoling thought is that he had before his death already done enough to make his name.'[43] Rutherford wrote a glowing eulogy in *Nature* lamenting the loss of such a brilliant mind.[44] The war had become personal, and this message was reinforced when he learned that Bragg's second son had been killed in the same theatre.[45]

The possibility that Germany might develop new technologies continued to worry Rutherford, and he was not the only one to wonder what might happen if the central powers obtained a technical edge.[46] However, it took a surprising time for Britain to harness the strengths of its own scientific community. In Britain the push was led by the Royal Navy, whose political head at outbreak of the conflict was Winston Churchill, a technophile of mercurial temperament. So too was his opposite number, the redoubtable 74-year-old First Sea Lord, Admiral Sir John Fisher. Churchill was one of the main movers and shakers behind the development of tanks, which began as a clandestine Royal Navy project.[47] However, he and Fisher fell out over the stalled Gallipoli campaign in May 1915, resulting in both losing their positions.

By this time one of the main problems at sea was the submarine threat, which gained strength in early February 1915 when Germany initiated 'unrestricted' U-boat warfare. That was a dramatic decision. Agreements reached at The Hague in 1899 required submarines to surface and stop any civilian vessel, searching it for war materiel and combatants, while preserving the lives of civilians aboard. By contrast, 'unrestricted' warfare enabled U-boats to simply attack civilian vessels on sight. Into this was mixed the fact that the Royal Navy's warships had been under that same threat since the outbreak of war. The issue was given impetus when three British cruisers were sunk by a single U-boat attack in September 1914, with the loss of 1459 lives. Britain's survival relied on naval superiority, keeping German merchant fleets blockaded and their navy unable to land forces on British coasts, while simultaneously foiling German efforts to shut down Allied sea trade.

In June 1915 the new political head of the Admiralty, Arthur Balfour, proposed what became known as the Board of Invention and Research (BIR) to look into naval technologies, particularly the thorny problem of finding submarines underwater. The Board was also given the task of reviewing the many inventions being offered by helpful civilians as war-winning devices. The BIR paralleled similar groups being set up by the Ministry of Munitions and other government departments,[48] marking the transition of research and development into organised large-scale arrangements in which development was collaborative and, often, run by committee. This approach was already emerging by 1914, but was given impetus by the war.

Balfour approached Fisher, hoping he might chair the new

organisation despite his recent fall from grace. Fisher agreed. Lord Rayleigh gave him a list of names,[49] and Fisher picked what he described as 'super-eminent colleagues' to join him on the Board — J.J. Thomson; Fisher's long-standing friend and engineer Sir Charles Parsons; and the chemist Sir George Beilby.[50] They were supported by an advisory panel that included Rutherford, Lodge, Crookes, Bragg and Rayleigh. The new organisation was based in a building in Cockspur Street, London, and met for the first time on 19 July. They divided their work into six sections, each handled by a subcommittee: airships, submarines and wireless, naval construction, anti-aircraft equipment, ordnance and ammunition, and aircraft armament. Each member of the advisory panel was expected to serve on at least two subcommittees.[51]

Rutherford was not entirely pleased to have been shoulder-tapped: the work drew his attention away from his work on atoms. But the war called, and the work was interesting enough. Among other things the board was deluged with helpful ideas for war-winning inventions — over 40,000 by mid-1917, including 14,655 involved with submarines, submarine detection and mines — all of which had to be assessed.[52] That joined a determined effort to explore new technologies, in which Rutherford's focus was underwater sound detection.

The fact that Rutherford switched from particle physics to underwater sonics has raised eyebrows among some of his biographers. Superficially these seemed to be different areas — how could someone who knew about one field somehow manage another? In fact the scientific methodologies applied across the board, as did the principles of physics. Indeed, Rutherford was able to co-opt equipment he had used for atomic research. And he was in his element, the intersection between technology and physics, working with long-standing colleagues such as Bragg.

Finding submarines underwater was a tricky business. The three approaches explored by Rutherford and his colleagues — magnetic detection, passive hydrophone direction-finding, and active echo-detection — were obvious ways forward. The problem in 1915 was that the physics needed better defining, and the results then had to be translated into practical apparatus. The Royal Navy had already dabbled with this principle, but it needed proper work. Rutherford took it on. The BIR provided funds, and he assembled a small team of former students and had a water tank installed in his Manchester laboratory.

The main problem was one Rutherford had tackled many times: detecting something that stood at the limit of available instrumentation. Faint underwater sounds produced only very tiny responses in underwater microphones, and the question was how to turn these infinitesimal vibrations into an electrical signal. Conceptually this was little different from the challenges Rutherford had faced when detecting electromagnetic fields in his early 'wireless' days. This time he thought he might exploit piezoelectricity, a phenomenon by which quartz crystals produced an electric current when squeezed. Jacques and Pierre Curie had devised an instrument for detecting piezoelectric current in 1882, largely forgotten to the field but known to Rutherford, who had seen it demonstrated in 1910 during the experiments required to establish a standard measure of radioactivity. He used this as a starting point and began a lengthy series of hands-on experiments.[53] He sent the results to Bragg who was working at a full-scale research station at Hawkcraig, where equipment could be tested at sea.

This work was interspersed with meetings in London, and early in 1916 Rutherford confessed to Boltwood that he was getting 'rather tired of long railway journeys in winter weather.'[54] He stayed with the Bragg family while in the capital, finding that Bragg and his wife remained 'very cut up' at the loss of their son.[55] One of the key challenges Rutherford had to solve was determining the direction of any incoming sound, something that was tricky, but perhaps less of a problem than picking up that sound in the first place. He drew on colleagues for support, including Louis King, the Canadian physicist who became Rutherford's successor at McGill, and who was working on ways of measuring 'sound intensity in absolute units', reducing those to 'electrical measurements.'[56] Rutherford and Bragg together developed a workable system, and in August 1916 the two scientists took out a joint patent for a directional underwater hydrophone.[57] Their efforts drew the attention of the Royal Navy: the Commander in Chief of Britain's Grand Fleet, Admiral Sir John Jellicoe, considered these underwater detection systems 'by far the most important matter at present before the BIR.'[58]

The BIR was not alone in these efforts. A parallel push was led across the channel by Rutherford's long-standing friend and colleague Paul Langevin, with the support of Constantin Chilowsky. Rutherford was aware of their work. The science historian Shaul Katzir has argued that this was what led the British to start investigating active detection by broadcasting sound underwater at frequencies well above the human audible range, then waiting for the echo.[59] Initially the BIR called it 'Supersonics' — as in 'super' sonics, not 'faster than sound' — but for security reasons the method was eventually dubbed ASDIC in an effort to hide its nature. The term apparently referred to the navy's Anti-Submarine Division with 'ics' added.[60] As so often with technology the development was not unique: Langevin and his team were working on the same thing at the same time and arguably were further along in development.

Rutherford's focus remained underwater hydrophones and applying piezoelectric techniques to the detectors.[61] This work also applied to ASDIC, where the same method was used to detect the echo. To this extent he was involved in the origins of what ultimately became known as sonar. But it was not solely Rutherford's invention, and it would have been surprising if it were. Like physics, engineering had reached the point where no one individual could stand behind every new development. Even the part Rutherford played in developing piezoelectric detectors was paralleled by Langevin's effort in France, similarly springboarding from Pierre Curie's original invention. One of the outcomes was a solid technical basis from which practical instruments could be developed, including the medical ultrasound machines routinely used for diagnostic work today.

Work for the BIR put Rutherford under considerable pressure, complicated by the fact that any system had to be capable of mass

Niels Bohr, c. 1922.

Library of Congress LC-B2-5894-8, Bain News Service, George Grantham Bain Collection

During what turned out to be his final visit to New Zealand in 1925, Rutherford visited his old childhood home, including his old school at Havelock. Here he is seen standing outside the school fence.

PAColl-2524, photographer Claude Mills, Alexander Turnbull Library, Wellington

Ernest Rutherford and Mary Rutherford, c. 1930.

1/2-044322-F, Alexander Turnbull Library, Wellington

In 1992 Ernest Rutherford was selected to feature on the New Zealand hundred dollar note, along with his Nobel Prize. The portrait selected shows Rutherford during his time at Manchester and was used on three successive note series.

Reserve Bank of New Zealand

The old Cavendish laboratory in 2024.

Lemuel Lyes

Rutherford's legacy was commemorated by the New Zealand-founded company Rocket Lab, led by engineer Sir Peter Beck. The company's Electron rocket, designed for the small-satellite orbital market, is propelled by nine engines bearing Rutherford's name, each producing 24 kilonewtons thrust at sea level. A later and larger rocket followed the subatomic theme with the name Neutron, the particle Rutherford predicted in 1920.

Rocket Lab, Trevor Mahlman

The nine Rutherford engines at the base of an Electron rocket first stage.

Rocket Lab

Statue of Rutherford as a child at the Ernest Rutherford Memorial, located near his birthplace at Brightwater, Nelson.

Daan Heeneman

production. Bragg warned Rutherford that the Admiralty wanted at least a thousand hydrophones.[62] The need to get practical systems to sea intensified in January 1917 when the Germans reintroduced unrestricted submarine warfare. Britain struggled to keep up with merchant losses that threatened to both starve the country — which relied on imported food — and to hamper access to the continent.

Rutherford's pre-war work with atoms largely fell by the wayside during these difficult years. He continued to develop Marsden's experiments as time allowed, intending to identify the long-range particle Marsden had observed. And he dabbled in other work. But everything had to be fitted around obligations to the BIR, and early in 1916 the Ministry of Munitions of War began leaning on universities to brief them on any scientific work that 'has not been directly authorised' — in other words, to abandon anything except what was required directly for the war.[63]

The whole pre-war physics community was affected in much the same way. Marie Curie set up mobile X-ray units to assist diagnostic work for soldiers wounded on the battlefield. Langevin, like Rutherford, focused on war technology. Universities were stripped: King reported from Canada that life at McGill was 'very dull just now with so few students and Gray and Eve away'.[64] Inevitably the physicists drawn into the military found even less of interest. 'A military life makes few demands on the intellectual side,' George Kaye complained to Rutherford in October 1916, while serving in the Coalhouse Fort in East Tilbury, 'and in despair I have been hunting up some old results of mine on the X-rays from different metals.'[65]

One of the few physicists able to continue their civilian activities was Albert Einstein, an avowed pacifist now working at the Kaiser Wilhelm Institute of Physics in Berlin. By the time war broke out Einstein had spent years figuring out how his concepts of relativity applied to gravity. By late 1915 the threads were coming together for what became a major paradigm-shifting contribution to physics, displacing Newton's gravitation and providing a new 'operating system' for the observable universe.

EINSTEIN'S THEORY OF GENERAL RELATIVITY

ALBERT EINSTEIN'S THEORY of General Relativity is a cornerstone of modern physics, describing one of the main 'operating systems' of the universe. With one brilliant concept and an elegant series of equations, Einstein was able to explain gravity and as a result also explain the whole of space and time.

Einstein's idea was surprisingly simple. Back in 1905, he had shown that space and time were interlinked and that mass and energy were flip-sides of the same thing. He now argued that four-dimensional space-time could be warped by mass-energy. Particles normally travelled in straight lines, but warped space-time pushed them into curved paths. Because a body resists any effort to change its speed and direction, the result was an apparent force that seemed to be pulling towards the centre of the space-time distortion. This is what we call gravity. But it wasn't a real force like electromagnetism. Instead, Einstein argued, gravity was a second-order outcome — specifically, a product of the conservation of momentum, making gravity an incidental byproduct of the basic structure of space-time and its response to mass-energy.

Einstein's first major challenge was finding a way of describing this mathematically, which he did not solve even in principle until 1912 when Marcel Grossman (1878–1936) introduced him to tensor calculus. This had been invented by Bernhard Riemann and others as a way of describing three-dimensional curved objects. The equations appeared simple, but once fully expanded were complex and time-consuming to work out.

That affected Einstein's second challenge — proving his hypothesis. The result was that in his initial papers he focused on three problems he thought could test his theory, including the way the Sun's mass and emitted energy would distort space-time. This, Einstein proposed, would bend the light of stars observed near its visible edge during a total eclipse. Matching that against their position at other times would reveal the degree of space-time distortion. He also showed, approximately, how his theory explained the precession of Mercury's orbit. This moved in ways Newton's equations couldn't explain. Einstein hypothesised that, thanks to mass-energy equivalence, the energy output of the Sun itself distorted space-time, adding to the distortions produced by the Sun's mass. It was this extra twist that pulled the perihelion point of Mercury's orbit along. Actually solving the equations to prove it was another matter: Einstein came up with an approximate solution.

Einstein presented his theory in four lectures to the Prussian Academy of Sciences during November 1915,[66] publishing revised papers in the *Annalen der Physik* early the following year. In the interim Karl Schwarzchild — who was then serving on the Russian front — came up with a more accurate solution of Einstein's equations for Mercury's orbit.[67]

Einstein's theory did not make an immediate impact: such things were of little import when newspapers were filled with war-casualty lists, and where the attention of science was on technologies that might win the conflict. Then, in 1919, observations by the British astronomer Arthur Eddington showed that starlight was bent. Einstein had been a rising figure in the field since the first Solvay conference. Suddenly he was catapulted to superstar status — and controversy because, inevitably, not everybody agreed with what at the time presented as a wholly radical idea.

EDDINGTON'S PROOF OF EINSTEIN'S THEORY OF GENERAL RELATIVITY

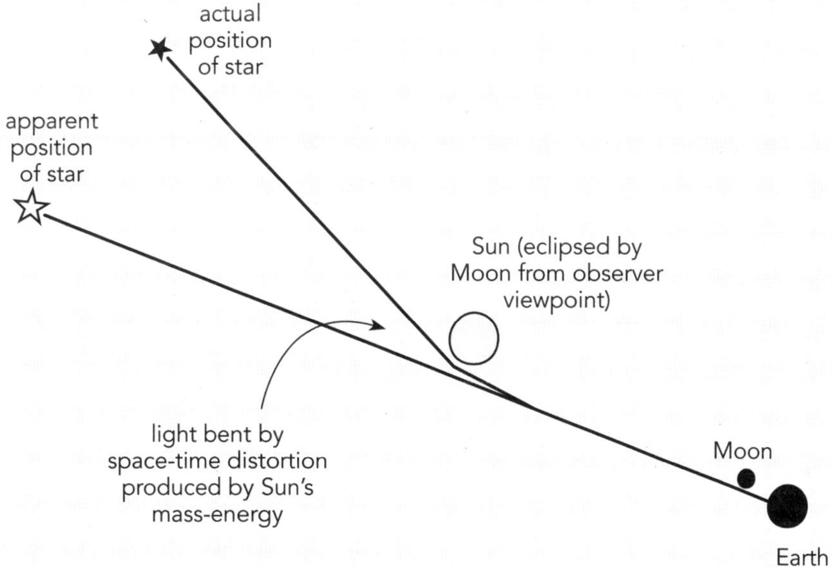

The mathematics of Einstein's General Relativity have since been shown to work every time, including for phenomena not known in 1915. This includes the expansion of space-time from a single point, the 'Big Bang' theory, proposed in 1927 by Catholic priest and astronomer Georges Lemaître. Another prediction of Einstein's equations was that if enough mass-energy was concentrated into a small enough volume, space and time would be warped around on itself, creating a 'black hole'. The solutions were worked out by the New Zealand mathematician Roy Kerr in the early 1960s, and the first black hole, Cygnus X-1, was identified in 1971. The first photograph of a black hole — actually of the disc of hot gas surrounding it — was taken in 2017 and released to the public two years later.[68]

Yet another prediction was that a large enough release of energy — the kind produced by two black holes colliding — should produce ripples in space-time called gravity waves. The technology to detect these took nearly a century to emerge. Efforts to detect gravity waves began in 2004, and in 2015 instruments built by the National Science Foundation recorded them for the first time, exactly as Einstein's theory predicted. Another prediction of general relativity was proved in 2023 when physicists at CERN were able to produce enough antimatter — specifically, anti-hydrogen — to run drop-tests. There was some speculation that antimatter might fall upwards because its charge and spin were the reverse of normal matter. Einstein's theory said no, that antimatter would fall normally, because gravity isn't a force and the curvature of space-time pushes all particles in the same direction. And he was right.

Today Einstein's theory of General Relativity is proven right millions of times daily, every time anybody uses GPS. The GPS system is accurate *solely* because it takes into account the distortions produced via both special and general relativity.

The United States entered the First World War in early April 1917, adding a dimension to Rutherford's work. The US Navy had been running a low-key liaison programme with the British for some time, operating from a small office in London. That kicked into high gear as the British began sharing important lessons learned from the application of the new physics. One priority was anti-submarine warfare: German U-boats were ravaging transatlantic shipping. In May, Rutherford was asked to lead a combined Franco-British scientific delegation to America, tasked with briefing scientists on British developments. It looked like a junket, so Rutherford made enquiries with Thomson: was the mission 'worth undertaking' and not purely 'of a complementary nature'?[69] Thomson thought it would be useful. Rutherford decided to go, writing to his mother on 15 May that 'you will be surprised to hear I am leaving in a few days for the U.S.A., via Paris.'[70] Rutherford and a Royal Navy representative, Commander Cyprian D.C. Bridge (1885–1939) went to France and met the French delegates, along with Langevin, Marie Curie and others. He had not seen his colleagues for some years and found Curie 'grey and worn and tired.'[71]

By 1 June the joint mission was located in the Hotel Powhatan on Pennsylvania Avenue in Washington, D.C., on a hectic schedule that included meeting Newton Baker, the US Secretary for War, and Josephus Daniels, the Secretary of the Navy. Three days of meetings followed with representatives of the Naval Advisory Board and the US Navy's Research Council, which Rutherford felt were 'highly successful.'[72] One of the American scientists, Irving Langmuir, later told Rutherford that the information had 'started our work on the submarine problem in the right direction' and had 'guided us throughout.'[73] Rutherford also had opportunity to renew connections with the US scientific community, including the physicist Robert Millikan (1868–1953), of Caltech in Pasadena, who had identified the charges on atoms in 1909 and did a good deal to support the mission. At the end of it Rutherford was able to whip north to Canada and spend two days in Montreal, where he visited his old haunts and contacted old friends, before returning to New York for the journey back to Britain. The liner was crowded with a US Marine unit whose mascot, a goat, decided to use Rutherford's cabin for its own.[74]

By this time the BIR was in trouble, largely due to Fisher. The old Admiral was a polarising character and the Royal Navy had long been split between pro- and anti-Fisher factions. His 1915 fall added fuel to the fires, and these politics spilled into the Board, giving it the nickname 'Board of Intrigue and Revenge'.[75] In September 1917 the new First Lord of the Admiralty, Sir Eric Geddis, disbanded the BIR on the basis that its relationship with the Admiralty was dysfunctional.[76]

Rutherford continued to contribute to military science, but now had the time to focus on his civilian work. This had been ticking along since early 1915 when he laid out a programme to investigate the long-range particle that Ernest Marsden had noted in 1913.[77] By this time Marsden had gone to New Zealand to take up a position as professor of physics at Victoria University, but Rutherford did not let the question go: it was another piece of the puzzle behind the nature of radioactivity, which had been the focus of the work he and his team had been pursuing for years. So he pressed ahead with the programme himself, though only intermittently until his military work slowed up.[78]

Rutherford knew that the process of radioactive decay involved fragmenting a nucleus, because one of the products — the alpha particle — was a helium nucleus. He suspected the particles Marsden and Geiger had found were hydrogen nuclei, and wanted to explore the possibility that these were also components of the nuclei of heavier gases. To do that he intended to fire alpha particles into various gases: hydrogen, helium, nitrogen and oxygen, and see what emerged if he was able to break their nuclei. The energies involved pushed the limit of what was possible with radium. Rutherford's equipment followed what had become a standard form for atomic studies, a box some 18 cm long, 6 cm deep and 2 cm wide with a space to put a small sample of radium that sat between the poles of a powerful electromagnet. The magnet was intended to force unwanted beta particles away from his phosphorescent scintillator screen. The box was airtight — and had to be exceptionally so in order to hold gases such as hydrogen, which could easily slip out of otherwise gas-tight spaces owing to the small size of its atoms.[79] A small opening at one end was covered by thin foils — silver, aluminium or iron — which enabled Rutherford to stop alpha particles but permitted heavier particles to punch through to a small phosphorescent screen. The resulting scintillations had to be manually counted.[80]

Rutherford ran the experiments himself with the help of his laboratory assistant, William Kay, who he got to help count scintillations. Much of the work seems to have been undertaken during 1918: his notebook for June, for instance, is filled with pages of equations calculating the way alpha particles moved through hydrogen and exploring different explanations.[81] Rutherford also consulted briefly with his former student Chadwick, who had been interned in Germany. Chadwick visited Rutherford soon after being repatriated in December 1918 and found the physicist 'engaged upon the artificial disintegration of nitrogen and other work.'[82]

War work still intruded. In early November 1918 he went to Paris 'on Admiralty business', crossing with the Admiralty delegation 'in the regular boats used by soldiers on leave ... Everyone had to wear a lifebelt.' They were 'escorted by destroyers and airships ('dirigibles'). He met Langevin and found the 'French in good spirits'. He found the Place de la Concorde filled with captured guns and went on to Lille, which had only been liberated a short while before, and which he found 'decorated with wreaths of flowers.'[83]

Rutherford did not publish his conclusions until 1919, when he published four articles in the June issue of *Philosophical Magazine*, taking up nearly three-quarters of the volume.[84] It was a significant breakthrough and a major waypoint in Rutherford's ongoing investigation of the reasons why radioactivity occurred. This had begun with his discovery that radioactivity was the result of natural breakdown in certain elements. Then he had identified the atomic structure responsible, the nucleus. Now he had shown that he could induce an artificial nuclear reaction by disintegrating that nucleus, and in the process confirmed the existence of a new subatomic particle, which he later dubbed the proton. It was an enormous step. The nucleus went from a mathematical entity to a measurable object, confirming the huge revolution that his 'empty atom' structure implied for both physics and chemistry.

Rutherford has been credited with transmuting nitrogen to oxygen during these experiments, making him the 'first' to transform matter — a point that has since been disputed.[85] But this question is the wrong one to ask. By splitting a nucleus Rutherford had, by definition, indeed artificially transmuted an element. This was the specific outcome of breaking apart an atom, and he said so in as many words in a 1933 paper, when he explained that he had achieved transmutation in 1919.[86] He said it again in 1937 in a speech he wrote for the Indian Science Congress, calling his 1917–18 effort a 'first experiment on the transmutation of nitrogen.'[87] But he never claimed he had transformed it into oxygen — indeed, he had no means of knowing whether he had, as his instruments in 1919 were not intended to detect the transformation product.

More crucially, Rutherford never made the atom-splitting claim in any of his four 1919 papers. Why? Primarily it was because this was not the purpose of his experiment — he had not set it up to get any empirical data of this nature, and his first speculations along those lines did not come until his 1920 Bakerian lecture, as we shall explore in the next chapter. Indeed, he did not even name the new particle he had produced, other than to note that it was probably a hydrogen nucleus. But he did, carefully, make clear that the alpha particles must have fragmented the nitrogen atoms. Doing more, he observed, was going to take more power. Just how much more wasn't clear: the way forces worked at such small distances had yet to be fully analysed and without that information he couldn't precisely estimate the energy needed to break an alpha particle free of a nucleus.[88]

Rutherford saw his wartime experiments as yet another step towards greater insight: but to him his work raised as many questions as it solved. The crucial issue from his perspective was the fact that he had induced an artificial nuclear reaction in stable material and confirmed a new particle that clearly came from the nucleus. These were paradigm-shifting discoveries, confirming his model of nuclear structure and endorsing the concepts he had proposed to date.

Rutherford's discovery of the proton came against a backdrop of a changed world. Even as he prepared his results for publication he had to confront the way that the war had overturned the scientific community. His Manchester team had been shattered by war. Germany remained an enemy from which the allies wanted to exact reparations. The chances of quickly returning to the easy pre-war world of collegial research across the continent seemed slim. Rutherford's war work came to a close in February 1919, marked by a formal dinner at the Midland Hotel in Manchester.[89] In a letter of thanks to the Admiralty he and his colleagues urged the Royal Navy to accept that 'naval war in particular tends to become more and more scientific in character,' and to 'keep in touch with scientific development throughout the world.'[90]

10 | The crocodile

The majority of the experiments were really started by his direct or indirect suggestion ...

— Charles Ellis[1]

DURING THE 1920s THE LAST MAJOR PIECES of Rutherford's atomic structure and radioactive puzzle came together, largely explaining why some elements were stable and others were not. Much of that was a direct result of Rutherford's own drive. In the 1920s the Russian physicist Peter Kapitza referred to Ernest Rutherford as a 'crocodile', a term he later declared, politely, was symbolic of a family father and of the way crocodiles push forwards without looking back, as Rutherford did when tackling science.[2] Impolitely, the nickname also described Kapitza's experiences when working under Rutherford. As at McGill and Manchester, Rutherford's quick temper and intolerance of both slackness and rule breaking — particularly those who defied his ban on smoking in laboratory areas — were a volatile combination. Kapitza admitted to being 'a little afraid' of Rutherford.'[3] Both meanings are likely true — and there is no question about the way Rutherford pushed the Cavendish forwards, orchestrating its direction according to lists he made of intended researches.

In this way Rutherford made a further major contribution to science, this time as both composer and conductor, drawing the very best out of those he directed. His star students and associates became, themselves, major figures in the field: John Cockcroft, Patrick Blackett and Kapitza among them. All of this was noticed. The US physicist Irving Langmuir visited in 1927 and was allowed to poke about, telling Rutherford afterwards that the 'work going on at the Cavendish is most impressive — it gave me great pleasure to go around alone ... and talk to the different men about their work — such business is refreshing and stimulating.'[4]

In many ways Rutherford's return to the Cavendish completed the circle: he had begun his career there under Thomson in 1895. And it was thanks to Thomson that he returned. Late in 1918 Thomson announced his intention to focus on his position as Master of Trinity College and to retire from his post at the Cavendish. The top job in the country suddenly became available, and Rutherford was an obvious contender. From the university's perspective his appeal was his solid track record in both running a department and carrying out original research. The Cavendish was going to grow — this was obvious from the wartime experience — and somebody was going to have to both oversee that and then weld the expanded organisation into a new whole.

Rutherford wasn't sure he wanted it. One challenge was that Thomson didn't want to let go of his own laboratory space, or his assistant and research students. Rutherford consulted William Pope, who thought he should still take the post. 'The amount of freedom which is allowed a Professor here cannot be realised by anyone who has not been in such a position.'[5] Then he turned to Schuster. 'I find myself oscillating a good deal between the relative certainties of Manchester [and] the advantages of Cambridge as a place for advancing physics.'[6] Discussions continued through March 1919, in part with Sir Joseph Larmor, then a member of the Council of St John's College. Rutherford discussed salary — which had to at least match his Manchester figure. Rutherford also tackled Thomson over laboratory space, though Larmor told him that 'your position in no case depends on it.'[7]

The outcome was that Rutherford was appointed both to the chair and the directorship of the Cavendish on 2 April, though he had to work the final term at Manchester. Thomson was made professor of physics and given space to continue his work. Rutherford told his mother on 7 April that it had been a 'difficult decision' to leave Manchester, but 'probably best' as the Cavendish was the 'chief physics chair in the country.'[8] Congratulations poured in: Nagaoka wrote from Tokyo fizzing at the 'happy news' that 'you were appointed Cavendish professor of physics, for which I send you hearty greetings from the Far East.'[9]

Rutherford's Manchester colleagues were sorry to see him go. Henry Miers, the vice chancellor, asked Rutherford to assess the future direction of the Manchester physics department. Rutherford provided one, though his drafts, with extensive crossouts and emendations, make clear he was struggling to get his thoughts together. Even his final version added words in afterthought.[10] In May the Senate passed a resolution recording their appreciation of Rutherford's work for the university, a 'series of brilliant investigations, carried out by himself and by others working under his inspiration', which had given his laboratory 'an historic association with some of the notable landmarks of physical science.'[11] Miers passed the endorsement on 'with mixed feelings', agreeing with the sentiments but regretting that he was losing Rutherford's advice and co-operation in the university.[12]

Cambridge beckoned. Mary found a two-storey home in Cambridge with wide grounds, Newnham Cottage on Queen's Road. They signed a lease and settled in. The house needed renovation, something still under way by early 1920. Mary forbade Rutherford to smoke anywhere except his study, did not allow alcohol on the property, and took charge of the garden. Eileen Rutherford was in poor health and unable to complete university entrance exams, but was allowed to begin at Newnham College.[13] Rutherford had a letterhead made featuring his three-line address in raised lettering: Newnham Cottage, Queen's Road, Cambridge.[14] The unadorned simplicity was typical of many colleagues but, in its plainness also captured the fact that he remained grounded.

For Rutherford the move was a step up, though it came at cost of time for research. He had a large administrative load and was increasingly involved in national scientific institutions — he became president of the British Association for the Advancement of Science from 1923, president of the Royal Society from 1925, and chaired the Advisory Council that stood behind Britain's Department of Scientific and Industrial Research from 1930. On the other hand, control of the laboratory gave him the opportunity to direct a large body of work in which he was the guide, mentor and innovator.

Just who might come aboard was another question. Rutherford wanted a mathematical physicist in the team. James Jeans urged him to hire Einstein, telling Rutherford in 1920 that Einstein was thinking of leaving Berlin and would 'probably consider an English offer.' He was, Jeans urged, 'just the man needed, in conjunction with yourself, to re-establish a school for theoretical physics in Cambridge.'[15] It was not wholly a pipe dream. Arthur Eddington's proofs of general relativity provoked controversy in the German physics community and Einstein was confronted by rising opposition through 1920, including a major anti-relativity rally in August that filled the Berlin Philharmonic auditorium. Jeans was not the only one to suggest that Rutherford and Einstein should work together, but it never came to pass.

Rutherford arrived at the helm of the Cavendish just as Western physics entered a new phase, a transition from small- to large-scale institutional research in which multiple teams were harnessed to tackle problems that required increasingly detailed study. The Second Industrial Revolution was in full swing, with its mass production, its factories, and its demand for chemists, physicists and engineers to support the growing edifice and feed constant demand for development and to monitor product quality. The trend had already been evident before the First World War but was accelerated by military needs and maintained momentum afterwards. This was a new age of big science, an era of detailed work in large laboratories staffed by dozens of scientists, often set up to support specific aspects of industry. Rutherford officiated at the opening ceremonies for some of these in Britain during the inter-war period, remarking to the Australian-born physicist Coleridge Farr in 1936 that British scientific institutions had expanded by 20–30 percent in the prior two years. They were increasingly well-funded: as Rutherford explained, the Cotton Research Institution alone had a budget of £70,000.[16] That was no small sum in 1930s money.

Industries had little trouble finding scientists to hire: this was the golden age of science, the era when wireless burst across the domestic world, when aircraft were taking to the skies in increasing numbers. Ambitious students flocked to the universities eager to enter the growing and exciting world of science, knowing jobs were out there when they graduated. The Cavendish ran headlong into this issue soon after Rutherford took the helm. He already thought the laboratory too small. Larmor agreed: 'Extension of the laboratory is urgent and under your drive will go ahead as soon as prices are tolerable.'[17] But the number of students rose faster than funds and plans could progress, growing to around 600 — including 60 naval officers — in spaces that, as Rutherford put it, were 'adequately equipped for little more than half that number.' It was not going to get better any time soon: Rutherford anticipated 'great demand for physicists ... trained in research methods.' His solution was to ask for new buildings and laboratory space along with 'the endowment of another Chair of Physics in the University.'[18]

All of this was a backdrop to Rutherford's main priority: pushing into the mysteries of atomic structure. By the end of the First World War the new electrodynamic paradigm was well established and the first forays were being made into the concepts of quantum mechanics that underpinned it. In point of fact — although those of the day did not perhaps fully realise it — many of the foundations of what we now call modern physics had already been laid. Newton's thinking had been recast. The first subatomic particles had been found — some were, indeed, being harnessed as research tools. Rutherford had explained radioactivity and then identified the basic structure of the atom with nucleus and surrounding electrons. Einstein had offered a non-Euclidian way of looking at space, time, mass and energy. But much remained to be discovered. There was growing evidence that quantum mechanics, the emerging 'operating system' that underpinned electrodynamics and which explained atomic structure, was more complex than anybody had yet considered.

Rutherford plunged back into the social world of Cambridge, which was wrapped in traditions such as the Cavendish Society annual dinner, a riotous event involving food, toasts and silly songs. Lyrics held in Rutherford's papers include a hymn to Einstein, penned by the Irish physicist Alfred Robb (1873–1936) and sung to the tune of 'Deutschland Über Alles.'[19] A programme for a Cavendish Society dinner in December 1920 gives a flavour of the formal events: it was signed by all 38 who went, with Thomson as special guest. The menu offered choices of soup — Clear Brunoise or Thick Duchess — fish in Mornay sauce, sweetbreads and curried prawns, roast goose with vegetables, mince pies, Queen Pudding, and more. Afterwards there were toasts to the King and another to the Cavendish, to which Rutherford responded. There was a song — 'Induced Activity' — and more toasts, then another song dedicated to J.J. Thomson.[20] It was all rather jolly, a return, at last, to pre-war jests and collegial fun.

Rutherford also rebuilt relationships with his European colleagues, picking up where they had all left off nearly five years earlier. The Dutch physicist Pieter Zeeman (1865–1943) wrote to congratulate him on 'the splendid results you obtained concerning the nuclear constitution of the atoms.'[21] Contact with Germany was still via intermediaries. He had not heard from Geiger since 1915,[22] but then, in May 1919 a letter arrived. 'I need hardly say that all what has happened these last four years had had no influence on my personal feeling to you and I hope, dear Prof. Rutherford, that you still take a little interest in your old pupil who keeps his years in Manchester always in pleasant memory.'[23] Geiger had never felt able to call Rutherford 'Papa', but the sentiment was clear enough.

Rutherford wrote to de Hevesy, via Bohr, filling him in on work and family news. 'I hope your health has not suffered during the war and that you will be able soon to get down to your own work again.'[24] He wrote to Alex Muller, then at the University of Geneva, inviting him to work at the Cavendish.[25] And he renewed contact with Meyer, who had been sending him papers through the war. Meyer reciprocated. 'It was a great feeling to get your letter ... after so long a time.'[26] Rutherford's French colleagues, too, were emerging from their war work, and he engaged Curie and others in vigorous correspondence discussing what they might call some of the new radioactive gases. It took a while, but within a few years much of the easy friendship of the pre-war years had recovered. 'Sehr geehrter Herr Kollege!' ('Dear Colleague!'), the Austrian physicist Felix Ehrenhaft declared by way of greeting Rutherford in a letter from 1924.[27]

Rutherford's main priority was picking up his researches where he had left off. This was not easy. Germany remained a major source of radioactive materials, but these were almost impossible to get in the post-war chaos. Late in 1919 Geiger warned that available production was being pre-sold months in advance and it was 'very unlikely' that the government would permit exports.[28] The other challenge was time. Rutherford found himself amidst a frenzy of work as he settled in to the Cavendish: as he told Marsden, he was having a 'very busy time' as he also had to look after research students.[29] This nonetheless enabled him to orchestrate their work in the direction he wanted. As lists of projected research he produced in the 1920s reveal, he was now very much a conductor, managing a wide range of scientific work and directing the whole in the direction he envisaged.[30] By this time Rutherford's main focus was inspired by Moseley's discovery of the relationship between charge on the nucleus and the atomic weight. This, he felt, was so crucial that every effort had to be made to discover it.[31] He brought a team on board to assist, including Chadwick and a young student that Chadwick had taught while interned, Charles Ellis. He also shoulder-tapped Patrick Blackett.

One of the most promising newcomers was a Russian, Pyotr Leonidovich Kapitza (Peter Kapitsa, 1894–1984). He had lost his wife and family during the influenza epidemic that followed the First World War. His plight drew the attention of friends who pointed him to the newly formed 'Commission of the Russian Academy of Sciences for renewing scientific relations with other countries.' The Soviet relationship with the west was virtually non-existent. However, in mid-1921 Kapitza was able to reach Britain with the help of German diplomatic intervention and in July he and the Soviet scientist Abraham Joffe (Abram Ioffee, 1880–1960) visited Cambridge, where Kapitza asked Rutherford if he could work there for a short while. Rutherford was reluctant, given the way the laboratory was already crowded, but accepted the request. So Kapitza joined the team of around 30 researchers,[32] initially working on the top floor under Chadwick.

Rutherford was wary at first of his new Russian colleague — in a rare display of politics he insisted that communism had no place in his laboratory or its reports. But he became increasingly impressed with Kapitza's abilities and let him stay on. In January 1923 Kapitza was allowed to begin research on a doctorate which, thanks to a backdating clause and recognition of his prior work in Russia, meant he was able to complete the degree the same year. He deeply admired Rutherford and began joking with him — something few others risked doing.[33]

By the middle of the decade their association was bordering on easy familiarity. When Kapitza was best man at Chadwick's wedding he sent photographs to Rutherford, who was then in New Zealand.

Rutherford responded promptly: 'I had a real good laugh when I saw the spick and span appearance of yourself and Chadwick ... I understand from Fowler that his top hat completed your toilet. If you went to work in your laboratory in the same attire you would create a great sensation.'[34] In 1927, Kapitza went to Paris and married Anna Krylova without telling Rutherford, finally writing to Rutherford from his honeymoon: 'I hope to be in Cambridge at work next Monday to share my love between my wife and the laboratory.'[35]

PETER KAPITZA'S 'SHORT CIRCUIT' MACHINE

DURING HIS FIRST years at the Cavendish, Peter Kapitza developed ways of exploiting a long-known phenomenon; the fact that the *rate of change* in an electromagnetic field, itself, produced a sudden spike of energy, known as the 'jolt' or 'jerk'. By shorting out a heavy electromagnet, Kapitza could briefly induce extraordinarily high magnetic field intensity, and began devising ways this could be used to study particles.[36] Early results were promising: by late 1925 he had a 'short circuit' machine to hand, built by Metropolitan Vickers with a circuit breaker of Kapitza's own design, which was initially able to generate brief magnetic fields of up to 270,000 gauss from a jolt current of 13,500 amperes at 1000 volts. For comparison, a standard two kilowatt fan heater draws about 8.6 amperes at 230 volts on a continuous basis. There was only one problem. When Kapitza decided it was time to increase the power there was a sudden explosion as the apparatus burst one of its coils. To everyone's surprise there was no other damage, although the noise was spectacular. Rutherford was on his way to Australia just then, so Kapitza wrote to him via P&O, explaining with dry humour that an arc of this power was safe for an experimenter 'if he is sufficiently far away'.[37] Rutherford was impressed with Kapitza's abilities and results.[38]

The issue of second-order jolt effects is why power spikes occur in domestic power supply. If supply is interrupted and restored, or if additional sudden loads are placed across the grid by switching, a secondary voltage is induced in the system that can be enough to burn out delicate electronic equipment. That is why surge protectors are essential. The principle involved is identical to the one Kapitza used to develop his short circuit machine.

A good deal of work in the Cavendish during the early 1920s revolved around Rutherford's interest in matter transmutation and alpha particles. Here he saw potential in an invention by the Scottish physicist C.T.R. Wilson (1869–1959): a smoke chamber that made it possible to photograph trails left by speeding subatomic particles, capturing their collisions. A Japanese doctoral graduate, Takeo Shimuzu, was working in the Cavendish and Rutherford gave him the task of modifying the smoke chamber for the experiments now in hand.[39] The main challenge was creating a system to take the many photographs needed to detect a rare collision of an alpha particle with a nitrogen nucleus. Shimuzu came up with a reciprocating piston linked to a movie camera, but had not completed the work before he had to return to Japan. Rutherford then put Blackett on to the job,[40] with input from Kapitza.[41]

As one of Britain's most prominent physicists, Rutherford was leaned on to present the latest discoveries, and began using lectures to various institutions to announce his ideas, much as he had used *Nature* 20-odd years earlier to publish the essential concepts of his work ahead of the Curies. One of his first major post-war lectures — and the one that essentially laid out his own agenda — was his Bakerian lecture of 3 June 1920. This was no small deal. The Bakerian lectures were a prestigious annual series by the Royal Society that began at the bequest of Henry Baker (1698–1774) and which by 1920 had become an essential part of the Society calendar. The onus was also on whoever spoke to say something meaningful, relevant and current. And Rutherford did not disappoint. Andrade, later, described his Bakerian lecture as 'one of the remarkable pieces of foresight in the history of science.'[42] In a few poignant words Rutherford pushed to the edge of the known subatomic world. Moseley's atomic numbers, he declared, were a new way of classifying the elements because they showed that the electric charge on the nucleus was 'of more fundamental importance than its atomic weight.'[43] This was a new concept, one that Rutherford championed and which eventually became mainstreamed. Today, elements are listed in the periodic table in terms of that charge on the nucleus, the atomic number.

In the middle of his talk Rutherford dropped in a few lines that made clear just how extraordinary his insight into the nucleus really was. From the experimental data to date, he explained, it was possible that an unknown particle could exist in the nucleus: a particle of zero charge that he speculated might be comprised of a hydrogen nucleus and electron in combination. He called it a 'neutral doublet' and suggested that such a particle would have hitherto unknown properties. Its zero charge meant it could move through both matter and atoms without hindrance. Once inside an atom, he speculated, such a particle might then form part of the nucleus, or be torn apart. He announced his intention to find it.[44]

What he had done was predict a wholly new subatomic particle, neutrally charged, solely from the very scrappy data to hand. And he expected to find it in the nucleus. Neutrons, when later found at the Cavendish, turned out to have the exact properties Rutherford expected, notably their ability to penetrate atoms with ease.

Towards the end of this speech Rutherford also touched on transmutation, explaining that it was 'natural to inquire as to the nature of the residual atoms' after disintegration, and speculating on how it was possible to predict which element should result, based on atomic weights.[45] In short, he knew very well what he had done in the last year of the war. Others realised it too, increasingly widely after the lecture was published. 'The work of your Bakerian Lecture is great stuff,' US chemist and academic Herbert McCoy wrote from Chicago. 'It certainly looks as if you have demonstrated the splitting of atoms at will.'[46] Meyer was effusive, explaining that 'my bad English is not able to tell you my feelings of admiration.'[47]

Rutherford coined the term 'proton' in 1920, referring to the hydrogen nucleus he had smashed out of the nitrogen molecule in 1918. The term riffed on William Prout's name. Rutherford had already hinted at Prout's classical notions of hydrogen as a building block in his 1914 Hale lecture, and now apparently adopted the name at the suggestion of Darwin and others.[48] He first used the term in a public address to the British Association at Cardiff on 25 August.[49] His published papers, however, remained cautious. As late as 1922, when he published the results of experiments he and Chadwick had conducted into a variety of elements, Rutherford still referred to them as 'H' particles, which at the time meant they were the nucleus of a hydrogen atom.[50]

Hands-on science of his own became a rare commodity for Rutherford in his new role, but he pressed ahead with ideas when he had the time. Chadwick later called some of them 'damned silly experiments,' though admitting that they wouldn't have been if they had shown a positive result. But even a negative result was useful, and it seems clear Rutherford was trying out possibilities. Chadwick admitted that while Rutherford 'would talk in what seemed to be a rather stupid way,' it was always clear that 'those words were inadequate to express what was in his mind,' and 'the same thing would apply to some of these experiments.'[51]

But time for research became increasingly rare. Rutherford's status as the pre-eminent physics professor in Britain continued to bring demands on his time that drew him away from research. In 1921 he was elected professor of natural philosophy at the Royal Institution, requiring four lectures a year. But he was able to make suggestions to his students and get a good deal of research under way at their hands. He also began building bridges. When he heard that the Institute for Radium Research in Vienna was in financial trouble he asked the Royal Society for money, which he used to buy the radium lent to him pre-war by the Institute.[52] They were overjoyed: times were hard in Germany, its scientists still shunned. He continued to push Moseley's concept of atomic numbers again, in a lecture to the Institute of Metals in 1922. Both Rutherford and Einstein were also appointed to the Royal Netherlands Academy of Arts and Sciences in Amsterdam, where suggestions were again made that Einstein should join Rutherford at the Cavendish. This never came to pass.

Despite the growth of specialised centres it was still possible for one man to direct a general physics laboratory — to act as conductor, holding the whole idea in his own mind and providing a vision for his staff to detail. This largely became Rutherford's life during the post-war years. As head of the Cavendish Rutherford had undiminished curiosity and energy and, as Charles Ellis later put it, most of the experiments were his own ideas.[53] Indeed, Rutherford actively planned what was happening. His notes from 1925, handwritten on University of Manchester letterhead paper, list off tasks against researchers. Other notes, written in pencil across multiple pages, project the direction of forward investigations.[54] In short, Rutherford was organising everything — writing the score and then conducting the orchestra. He kept tabs on the results, monitoring everybody's work, offering suggestions, guidance and direction.

Rutherford's main focus remained the secrets of the nucleus, and he gave the majority of that work to Blackett, who picked up on the new 'H' particle where Rutherford had left off. Unknowns included the exact mechanisms by which the particle was ejected, and identifying just what was left after an atom of a known element was broken apart. The modified Wilson smoke chamber was the ideal tool for the job, and Blackett spent the next four years working up results from a lengthy succession of experiments that produced over 23,000 photographs. In the process Blackett was able to obtain empirical proof of matter transformation under alpha bombardment, finding the process was more complex than simply breaking apart atomic nuclei — but confirming, for the first time with hard data, that bombardment with alpha particles would cause nitrogen to break apart into a proton and an isotope of oxygen.[55]

In the northern summer of 1925 Rutherford made what turned out to be his last visit to his home country. It was at once poignant, personal and public. He knew his parents were ageing and wanted to spend time with them — possibly for the last time; but he made the journey as an extension of a lecture series to which he was invited in Australia. News that he was heading home sent a wave of enthusiasm through New Zealand. Percy Burbidge, who had been a student at the Cavendish and then became professor of physics at the University of Auckland, made sure Rutherford knew 'how glad I am you are coming,' declaring that it was a 'treat to get a physicist to talk to.'[56] Rutherford discovered arrangements were being made for him to attend civic receptions, give addresses and to deliver lectures.

Rutherford, Mary and Eileen left Liverpool on 3 July for a journey that included a fair amount of tourism.[57] While in Melbourne he caught up with his old high school mathematics teacher, William Littlejohn, for a 'very pleasant evening with him and some of his friends.'[58] By late September he was in Sydney, appearing to the media as a 'tall, clear-eyed florid giant with boyish personality absolutely radiating optimism ... six feet of cheerful New Zealander.'[59] Then he and Mary took passage on the *Niagara* to Auckland, intending to spend time with family before Rutherford embarked on a round of official events. The media nabbed him as he disembarked in Auckland.[60] Mary went on to Christchurch to see her family, while Rutherford spent a week in New Plymouth with his family.[61] Then he caught up with his brother Arthur at Waihou, near Hamilton.[62]

The official side of Rutherford's visit to New Zealand appeared as a triumphal procession. He was a national hero and everybody wanted to be part of the action. Rutherford wrote to Chadwick that he was being kept 'very busy with the various social engagements ... civic receptions, conversaziones, Governor's luncheons and talks to various bodies,' but that 'fortunately I keep very fit and have had a good time.'[63] In Wellington he was met by Marsden and a number of officials, ahead of delivering several lectures.[64] On 30 October he visited his old school in Nelson, where he addressed a 'gathering of boys and girls' and 'was accorded a great reception.' He took the opportunity of being in the district to visit Foxhill and his old school, where he was shown his name on the original register.[65] On 2 November he was in Christchurch, where 70 students surrounded his car on the Worcester Street bridge and 'with ropes dragged it to the Municipal Chambers' and a civic reception.[66] By 7 November he was in Dunedin, where students of Otago University gave a 'vigorous haka in his honour.'[67] The *Otago Daily Times*, slightly immodestly, described him as 'the world's greatest living physicist.'[68] From there Rutherford made a flying visit to New Plymouth to farewell his parents, and then he and Mary left on the *Maheno* for Sydney, where he had more lectures scheduled.[69]

It was Rutherford's final departure from the country of his birth. He did not see his parents again. His father died in 1928, leaving an estate worth £12,000 — around $1.48 million in mid-2020s values.[70] This was not too bad for a self-made man of the era, underscoring the fact that the Rutherford family were not poor: the financial issues confronting Ernest in his youth were in large part a product of instability and fluctuating income, not because the family were at the bottom of the socioeconomic pile.

At the end of November Rutherford was elected president of the Royal Society and awarded yet another honorary degree, a doctorate, by the University of Paris.[71] And so he returned to Britain on board the P&O steamer *Naldera*, via Suez, with further honours and awards under his belt.

By this time Rutherford's European colleagues were well down the track to explaining the concepts behind the subatomic action being explored at the Cavendish. Europe took time to get back to its feet after the war, but by the mid-1920s the physics communities in France and Germany were returning to a semblance of pre-war normality. The hot topic was the underlying mechanism of the subatomic world. The general 'operating system' was clearly related to the energy steps (quanta) identified during the immediate pre-war years, but by the 1920s that posed more questions than answers. Planck's 1900 argument that energy at subatomic level was stepped, for the obvious reason that particles were single 'packets', was only a beginning. Einstein's 1905 paper on the photoelectric effect had shown that these particles were also waves, creating the counterintuitive notion of 'wave-particle duality'. That was clearly going to affect the atomic model Rutherford and Bohr had come up with — which itself wasn't perfect, either. Rutherford's theory of the nucleus was clearly correct, as was Bohr's discovery that quantisation fixed the problem of electrons spiralling down to the centre. But that didn't account for the way that spectral lines from heavier atoms were split by powerful magnetic or electric fields. And then there was a need to encompass the issue of wave-particle duality, which was both real and — on the face of it — nonsensical.

The answers seemed likely to come from a better understanding of quantum mechanisms, and in wake of the First World War a new generation of European physicists began exploring the field. One breakthrough came in 1923 when the French physicist Louis de Broglie published three articles in *Comptes Rendus* that reconciled Einstein's photoelectric effect and special relativity with the Rutherford-Bohr atomic model. In this work, de Broglie was able to explain why electrons were quantised and showed that they manifested not as particles, but as standing waves. Wave theory — well explored mathematically in the early-to-mid-eighteenth century — made clear that such waves had to exist at discrete distances from the nucleus.[72] De Broglie's treatment was purely mathematical, but practical experiments by US physicists Clinton Davisson and Lester Germer at the Western Electric laboratories demonstrated that de Broglie's concept was correct. He received the Nobel Prize for it in 1929.

The sceptre was also picked up by the German physicist Werner Heisenberg, then working at the Institut für Theoretische Physik in Göttingen, who proposed a way of blending Newton's classical mechanics with quantum explanations for subatomic behaviours. His logic was straightforward: classical mechanics clearly works at macro-scale, but the macro-scale world is merely an assembly of subatomic particles and their behaviours. On that logic, classical mechanics as defined by Newton was a specific expression of more general rules that applied to subatomic particles. What happens, Heisenberg wondered, if those equations are inserted into the new paradigm?

This concept put quantum matters front and centre of physics and portrayed electrodynamics as an expression of more fundamental quantum mechanisms. The challenge then became reconciling everything in terms of the new paradigm. Heisenberg knew his paper was only a start,[73] but it did not take long for others to pick up the theme. He gave the paper to Max Born for publishing. Born read it and began worrying about one of Heisenberg's equations, realising a solution could be found through applying matrix mathematics. Born and his colleague Pascual Jordan then entered the fray with a paper of their own. Along the way Born coined the term 'quantum mechanics' to describe what was fast emerging as a new 'operating system' for subatomic behaviours.

What these theoreticians struggled with was the elusiveness of wave-particles. Heisenberg showed that uncertainties when identifying position and momentum were related to Planck's constant, meaning that both couldn't be defined simultaneously. In 1926, Erwin Schrödinger produced the quantum equivalent of Newton's Second Law of motion. Newton had shown that the acceleration of a moving object was a result of its mass and the forces acting upon that mass. Schrödinger applied the idea at quantum level, producing a way of describing what has become known as the 'wave function', which was defined as a range of probabilities. This was then adapted by Paul Dirac into an equation that included special relativity. The effort to make quantum mechanics apply to everything was gaining momentum. Rutherford was well aware of these developments: the physicists involved were all friends and colleagues who actively pointed out the papers to him.[74]

Just what all this meant was hammered out at the fifth Solvay Conference of 24–29 October 1927, again chaired by Hendrik Lorentz. This produced what remains the accepted explanation of how quantum mechanics work, although the explanations promoted by Bohr and Heisenberg at the conference were counterintuitive. The basic idea was that the subatomic particles that defined reality didn't exist in a 'real' sense until measured. Until then these particles were merely a 'wave-function', a range of probabilities that 'collapsed' into a single reality when any attempt was made to measure them. It was an attempt to explain and rationalise the fact that particles appeared as *both* waves and objects, and that any effort to observe one — in other words, to measure its velocity or momentum — relied on other particles such as photons, which also interacted with the target particle. A new vocabulary had to be devised to try and describe it. The mathematics worked out, however the concept was deeply abstract and wide open to misinterpretation. Einstein was convinced they had missed something, but couldn't come up with a compelling countercase.

Further proofs that quantum mechanics was real — irrespective of how physicists explained the matter — followed swiftly after the conference. Back in 1895 Rutherford had been introduced to J.J. Thomson's son George, then a shy three-year-old. George Paget Thomson followed in his father's footsteps and became a physicist. In late 1927 he used the standard double-slit experiment to show that electrons, like light, produced wave-interference patterns. This work came in parallel with similar tests being conducted in the Western Electric laboratories on the other side of the Atlantic by Davisson and Germer.

The model of the atom that emerged from this thinking was virtually identical to Rutherford and Bohr's, but now portrayed electrons as waves occupying probability zones around the nucleus, defined by energy steps. This did not mean earlier discoveries were wrong: on the contrary, Rutherford had correctly shown that atoms had a nucleus that occupied a tiny fraction of the volume and Bohr had correctly shown that quantised energy steps were required to make the electron orbits stable. Now a further layer was being added as additional information emerged about the quantum paradigm, revising the details along the way. It was a classic demonstration of the Western scientific method. And Rutherford's nucleus — minuscule, yet holding the bulk of the mass and charge of the atom — remained the key to the whole structure.

The explanation of quantum mechanics that was hammered out at the 1927 Solvay Conference subsequently became the way by which all quantum mechanics was explained, and around which it was eventually extended to an explanation of all physical phenomena, including electrodynamics, which became 'quantum electrodynamics' in the new paradigm. Quantum mechanics, in short, explained everything — everything, it turned out, except gravity. That was the sticking point. Quantum gravity required a particle to carry it. None was ever found, and the sole explanation for that phenomenon, stubbornly, remained Einstein's theory of General Relativity, which viewed gravity as a second-order outcome of warped space-time, not a real force in itself.

Ultimately the problem wasn't with what was happening at subatomic level, but with the way it was being interpreted by scientists operating within Europe's scientific framework and style of thinking. The determinism demanded by this framework — the need to have specific, measurable outcomes — coupled with the hard-edged categories into which observations were jammed, essentially pushed Bohr and Heisenberg down a particular line. It was self-consistent and difficult to refute, but it was also virtually impossible to grasp. The problem, ultimately, was not with the phenomena being observed, but with the way these observations were being received. Wave-particle duality crossed boundaries. Ultimately the problem was that Bohr and Heisenberg won an argument, squashing alternative ways of understanding the observations. Other ideas of the time included de Broglie's 'pilot wave' theory. Later, the British physicist Roger Penrose (1931–) showed that wave-function collapse could also be seen in Einstein's relativistic terms.

One point seldom made is that Heisenberg's framework also had another social overlay: the mood of post-First World War Weimar Germany. It was a nation struggling to come to terms with shattered Imperial dreams and the collapse that ended the First World War, a collapse given depth by runaway inflation and economic ruin as the Allies levied reparations. The resulting mood suffused German society, typified by the dark visions of film-maker Fritz Lang, and is still with us today in the legacy of Bauhaus architecture. Various odd philosophies gained ground at the time, along with extremist political movements — all reactions to the mood of the day. This same general social framework also suffused everyday life, a constant backdrop to the physicists trying to understand the world revealed to them by Rutherford and others.

The wider problem for physics was that the Bohr-Heisenberg concept, which Heisenberg later called the 'Copenhagen' interpretation, was almost incomprehensible. One outcome when quantum mechanics filtered through to the general public was that it became popularly considered akin to magic. There were even arguments to the effect that wave-function collapse occurred *only* when observed by human consciousness, which was deeply anthropocentric. Efforts that some physicists made to lampoon the Bohr-Heisenberg explanation — notably Schrödinger's cat — have since been misinterpreted as representing the principles. Rutherford himself played an indirect and likely unconscious part in this: Bohr was Rutherford's student who had spent years working under Rutherford with his focus on teaching students *how* to think.

These developments — the proof that quantum mechanics underpinned electrodynamics — are pivotal to understanding Rutherford's place in modern physics. The electrodynamic revolution of the 1890s had taken the classical physics of earlier times and recast them into the world of electromagnetic forces, creating a new way of understanding the very structure of space and time, and explaining mysteries such as the missing aether along the way. Now a new generation of physicists argued that those electrodynamic frameworks were themselves a second-order expression of even deeper processes involving the fundamental particles being discovered by Rutherford. It was, in effect, a process of delving into layers: each new layer was viewed as a way of understanding everything until the next layer was found. But this did not invalidate the discoveries that came with each layer — those had to be taken into account when understanding the next. Einstein had not invalidated Newton but instead extended him and, in the process, found a different explanation for the same phenomena. The same applied to the discoveries of electrodynamics when Bohr, Born, Heisenberg and the rest began investigating quantum mechanics.

The logic that everything came, ultimately, from particles and

SCHRÖDINGER'S CAT

THE PHYSICIST Erwin Schrödinger disagreed with the explanation for quantum mechanics that Niels Bohr and Werner Heisenberg pushed to their colleagues during the 1927 Solvay Conference. Schrödinger had devised an equation to describe the range of uncertainties about the position, mass and spin of particles that resulted from the fact that it behaved like a wave. This was known as the wave function. If some feature of the particle was measured, however — for example, if its mass was identified — then the wave function collapsed and certainty followed. Because an unmeasured particle could have a range of these things, Bohr and Heisenberg argued that an unmeasured particle had to have *all* of the possible values simultaneously — that it was in a state of 'superposition'.

Schrödinger thought the idea ludicrous, because if this principle applied at microscopic level then it should, logically, also apply to the macroscopic everyday world. To highlight the point he came up with a thought experiment. Imagine a cat sealed into a box with a glass phial of poison and a device to break the phial, triggered when a small quantity of radioactive material releases a particle. Such release is wholly random. Consequently, nobody could know whether the cat is alive or dead. By Bohr and Heisenberg's 'superposition' argument, in which every state exists simultaneously until measured, the cat is therefore *both dead and alive at the same time.*

Schrödinger made the point to demonstrate that this quantum explanation couldn't be extended to macroscopic level without creating absurd situations, which was a problem if quantum mechanics was meant to sit behind everything. The problem wasn't with quantum mechanics, but with the way it was being explained by Bohr and Heisenberg. To Schrödinger and a fair number of other physicists, including Einstein, there had to be another way of explaining how quantum mechanics worked. But they were unable to prevail against the Bohr-Heisenberg bulldozer.

Ironically, Schrödinger's cat later became a popular way of explaining how the 'Copenhagen' interpretation is meant to work.

their behaviours — for which the basic 'operating system' is quantum mechanics and electrodynamics only a secondary effect — remains a key assumption of modern physics today. Where did Rutherford stand in those terms? He had built his life's work around electrodynamics and determinism, using it to solve the mystery of radioactivity, discover the nuclear structure of atoms, and found the science of particle physics. He had pioneered and then dominated that work by applying skills he first developed at Canterbury University as a student. By the late 1920s much remained to be discovered about particles and the new explanations offered by Bohr and Heisenberg still made little practical difference to Rutherford's laboratory work. But they did intrude into the explanations he might offer for what he was observing.

RUTHERFORD'S NUCLEAR ATOM WITH QUANTISED ELECTRON CLOUDS

THE FINAL EVOLUTION of Rutherford's nuclear structure came with further developments in quantum mechanics during the late 1920s, in which Louis de Broglie and Erwin Schrödinger found that electrons existed as wave functions with indeterminate position within each energy level around the nucleus. By this time the proton was known and Rutherford had predicted the neutron, which was found just a few years later.

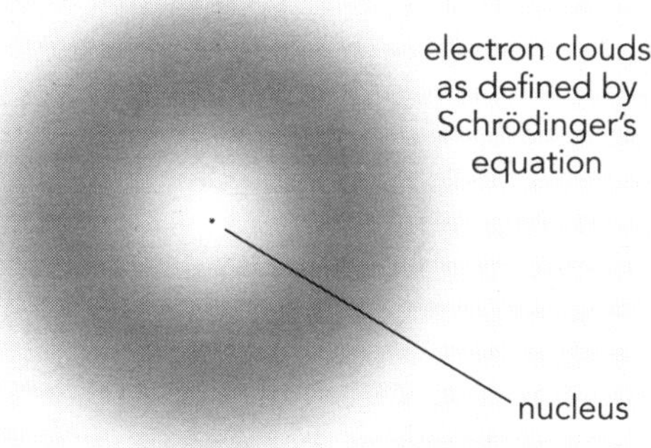

Where did Rutherford stand in regard to quantum mechanics? He had always been jocularly cynical towards theoretical physicists. He routinely worked himself with matters invisible to ordinary human senses, but these still registered on his instruments and to his practical mind the theorists were too airy, the ideas too abstract. The new quantum theories were little different. Associates of the day such as Charles Ellis found him either uninterested or of the opinion that quantum mechanics was not required for his own work.[75] But as always with Rutherford, it appears he took the theoretical ideas on board. A lecture he gave in March 1932 on the origins of gamma radiation made clear that he accepted quantum mechanics as the organising principle behind everything else.[76]

Whether Rutherford agreed with the Bohr-Heisenberg explanation of quantum mechanics is another matter. The logic that everything flowed from the behaviours of particles was sound, and the observations added up: quantum mechanics was real. But there were problems with the Bohr-Heisenberg explanations as to *how* all that worked. Einstein certainly had his doubts. The issue was similar to seventeenth century arguments over the aether. Newton thought the idea silly, but eventually aether became dogma and it took nearly 250 years to discover that Newton had been right after all to suggest light was carried by particles, though for reasons only found through further discoveries. In 1927, as the new paradigm of quantum mechanics exploded upon the world of physics, due cynicism about the *explanations* was justified.

The main problem that Rutherford and other experimentalists shared by the late 1920s was less to do with theory than the hard fact that long-standing methods for investigating the subatomic world were reaching their limits. Alpha particles released from radium carried phenomenal energy for their size, but as Rutherford noted in 1919, more power was going to be needed to overcome the Coulomb forces involved with the nucleus of the heavier elements. The substance of the nucleus itself — the proton — also had potential as a new research tool. Protons could be isolated artificially via the methods he had already devised, but energy was going to be needed to get them up to a velocity where they could be used as a research tool. Rutherford knew he was going to need this sort of power for his next goals — and he also knew that other physicists had come to the same conclusion. He and the Cavendish were, once again, on Tom Tiddler's ground.

Rutherford was keen to test his hypothesis that a neutrally-charged particle of similar mass to a proton existed inside the nucleus. That was likely to need particles with more energy than radium could provide. He put two of his staff on to the problem, Ernest Walton (1903–1995) and John Cockcroft (1897–1967). Once again the effort was something of a race. By the late 1920s efforts were under way to build machinery that could accelerate charged particles to enormous velocities. German physicist Max Steenbeck proposed a method in 1927, and the following year the Hungarian physicist Leó Szilárd came up with a design for a linear accelerator, closely followed by a circular device known as a cyclotron.

By this time the United States was a rising power where the sciences had considerably greater scale and financial backing than was available in post-war Britain, and the point evidently drew Rutherford's attention. He hosted a visit by Langmuir in late 1927 where discussions apparently revolved around André Siegfried's recent book *America Comes of Age: a French Analysis*.[77] This inspired Langmuir to engage in a fruitless search through Cambridge and London bookshops for a copy. He finally found one in French in Paris and read it on the way back to the United States, reporting to Rutherford that he found it 'extremely interesting' and the 'best analysis of American conditions that I have read.'[78] The study made clear what was happening: after decades of introspection and a focus on the Western frontier the United States was emerging into the world — and doing so on a scale given impetus by the relative place of post-war Europe.

The point was clear when it came to particle physics, where

Yale graduate Ernest Lawrence (1901–1958) came up with a new but inevitably expensive system for transcending the energy limits of radium. Lawrence was hired as associate professor of physics at the University of California, Berkeley, in 1928. The following year he learned of a method invented by the Norwegian engineer Rolf Wideröe by which particles could be accelerated twice by the same electrical potential, simply by reversing the polarity — in effect, a push-pull arrangement. Lawrence now wondered if it was possible to accelerate particles magnetically around a circular track — whether they could be accelerated continuously and so given enormous kinetic energy, a unit measured in 'electron volts'. He began working on a practical machine in 1930 with the help of fellow physicist Milton Stanley Livingston (1905–1986), who had approached him looking for a thesis topic.[79] By early 1931 they had built their first working cyclotron, which was able to drive protons up to energies of 80,000 electron volts. That became a proof-of-concept for a machine able to achieve particle energies of a million electron volts or more.

Rutherford had no intention of falling second to the Americans. The work that followed was led by Walton and Cockcroft, with input from the Australian physicist Mark Oliphant and others, but Rutherford stood closely behind it — in effect, orchestrating the whole. Rutherford estimated that energies of up to a million electron volts — and probably far more — would be needed to smash apart an atomic nucleus. The problem wasn't the physics so much as the finances involved in building equipment able to give protons the necessary kinetic energy. The financial challenge was solved through quantum mechanics. In 1928, the Soviet physicist Georgiy Gamov (George Gamow) gave a presentation to the Cavendish outlining how the wave function enabled particles to pass through the Coulomb repulsion forces of a nucleus at much lower energies than classical physics predicted. The phenomenon was described with an equation developed by Schrödinger and became known as 'quantum tunnelling'. Cockroft discussed the phenomenon with Gamov,[80] then ran some calculations and realised that this meant the energy required for

ELECTRON VOLTS AND CYCLOTRONS

THE ELECTRON VOLT, eV, is a measure of kinetic energy — that is, the energy of movement. An eV is defined today as the energy gained by a charge equivalent to one electron when it moves through an electric potential difference of one volt, in a vacuum. A single eV is a minuscule quantity of energy: in standard international units it is 1.6×10^{-19} joules — in other words, 0.00000000000000000016 joules. For comparison, a standard two-kilowatt fan heater produces 7300 joules an hour — the energy equivalent of 6.214×10^{18} electron volts.

Protons, the particles Rutherford isolated in 1918 and later harnessed to investigate nuclei, are 1836 times the mass of an electron and accordingly require more energy to accelerate. This is why Rutherford and his team talked about millions of electron volts when developing equipment to give streams of protons enough kinetic energy to overcome the repulsive Coulomb force of the nucleus. This was far more than could be provided by laboratory apparatus of the kind used at this time. The practical solution was found to be a circular accelerator, sometimes known as a cyclotron, which enabled a stream of particles to be accelerated around a spiral path, dramatically increasing the distance over which they could be accelerated by comparison with a linear track.

The cyclotron remains in use today for a variety of purposes, notably producing radioisotopes for medical imaging. However, its role as a research tool for physicists has largely been replaced by synchotrons such as the Large Hadron Collider (LHC) at CERN. These differ from cyclotrons because they accelerate protons continuously around a circular track, achieving considerably higher energies than a cyclotron. Inevitably such devices can be built only at tremendous cost, in the case of the LHC around €7.5 billion. Part of the reason for their scale is that these devices lose efficiency owing to synchrotron radiation, the energy bled off by high-speed particles when forced around a curved path. This phenomenon was predicted by Maxwell's original equations in the 1860s but not observed until 1947, at the synchrotron operated by the General Electric research laboratory in Schenectady, New York.

their project could be as low as 300,000 electron volts. According to Blackett's experiments, this would give protons a range of about five millimetres in air, just enough to be visible in a Wilson chamber.[81] The reduced scale put the required gear within reach of the shoestring budget Rutherford had available.

Cockroft and Walton began working on the equipment, essentially in parallel with efforts to do the same thing at Berkeley, where Lawrence had a far higher budget. One of the challenges was powering the rig, and in particular handling the very high current loadings the system required. Cockroft and Walton initially thought of using Kenotron rectifying valves developed by General Electric. Like all thermionic valves, these were yet another descendant of the Crookes tube.[82] The problem was that the system demanded multiple Kenotrons, which weren't cheap, and if one failed the rest would follow. That required considerable work to sort out. Then there was the issue of high-voltage supply, for which a custom 350 kilovolt transformer was eventually designed and built by the Metropolitan-Vickers Electrical Company.[83] Development was still iterative, delayed by the need to move the laboratory, along with a decision to increase the power.

While Walton and Cockroft worked on their particle accelerator, Chadwick got working on another problem: Rutherford's predicted neutral particle in the nucleus. By 1930 clues that such a particle might exist were steadily emerging. Walther Bothe and Herbert Becker discovered that remarkably intense radiation, unaffected by a magnetic field, was produced when alpha particles produced from polonium struck light metals such as beryllium. They hypothesised that it might be gamma radiation. Then, early in 1932, Marie Curie's daughter Irène Joliot-Curie and her husband Frédéric Joliot showed that this high-energy radiation was able to drive protons out of hydrogen molecules. It didn't seem to be gamma radiation, and Rutherford didn't think so either. The Cavendish had better equipment for obtaining granular data than the French team, and he put Chadwick on to the job.

Chadwick showed that particles radiated from beryllium were able to easily penetrate lead that stopped protons. These new particles had a mass virtually identical to the proton, but no electrical charge. They seemed to be the particles Rutherford had predicted in 1920, though Chadwick discarded the notion of a combined proton and electron, because quantum mechanics showed that the only possible combination of those particles was a hydrogen atom.[84] What he had found, then, was a brand new particle — the neutron, able to penetrate atoms with eerie efficiency just as Rutherford expected. Chadwick later received the Nobel Prize for the discovery. Research has since shown that in stellar cores a proton and electron can transform into a neutron and release a neutrino. This is part of the process of nuclear fusion, which is why the Sun shines. Elsewhere, neutrons can decay into a proton, electron and an anti-neutrino by a process known as $\beta-$ decay, which is the origin of beta radiation. Both are outcomes of the way weak force operates. This was not known in 1932, although Rutherford realised that forces other than electromagnetism had to be at work inside nuclei. As it happened, the discovery of the neutron set physicists on a path that led to weak force being identified.

Cockroft and Walton's particle accelerator now entered the fray. The device had aspects of a home-built lash-up, to the point where the operators shielded themselves from expected radiation inside a small wooden hut just big enough for an operator to sit inside, lined with lead. Inside they hung a screen coated in zinc sulphide: if it glowed during experiments it meant they needed to increase the shielding. From inside the hut it was possible to use a microscope to observe a scintillator screen. And on 14 April 1932, Walton ran an experiment that split lithium atoms with a stream of protons, dividing them into 'two α particles with a total energy release of 17.2 million volts'.[85] He called Rutherford and Cockroft in to check the results. They had done it, shattering an atomic nucleus in a particle accelerator for the first time. But this was only the beginning. Rutherford told Marsden he was 'very interested to see whether anything special is to be observed,' anticipating that 'if we get a union or capture we should expect to get out an α-particle of very great energy.'[86]

Rutherford, meanwhile, began work of his own with Oliphant to experiment with artificial transmutation using protons at energies of up to 200,000 electron volts — less than Cockroft and Walton were using but still higher than could have been obtained a few years earlier with radium.[87]

The next discovery from the Cavendish was just as extraordinary. In 1928, Paul Dirac published a paper explaining why spectral lines divided into components inside a static magnetic field, a phenomenon known as the Zeeman effect. To resolve this Dirac took quantum mechanical principles, added the concept of electron spin, and stirred in Einstein's Special Relativity. The resulting equations explained the effect, but also yielded a solution that seemed nonsensical: it predicted positively-charged electrons. None had ever been seen, but Dirac's equations indicated they were mathematically possible. The issue was batted back and forth among the physicists, including J. Robert Oppenheimer (1904–1967). Later, in 1931, Dirac extended Oppenheimer's ideas and came up with the answer. There had to be an anti-electron, identical to the one known by physicists for decades, but with opposite spin and electric charge. When an electron and its anti-particle met, they annihilated each other, releasing energy.[88]

By this time a number of anomalous results had been observed in particle experiments that — in hindsight — turned out to be the mysterious 'anti-electron' — but the first confirmed discovery did not come until mid-1932. Blackett was working with the Italian physicist Giuseppe Occhialini (1907–1993), using a Wilson cloud chamber to explore cosmic rays, mysterious radiation of enormous intensity that poured down on Earth from space. In the process the physicists discovered particle pairs that turned out to be an electron matched with its positive equivalent. They held off announcing it and lost out to US physicist Carl Anderson (1905–1991), who discovered the same thing in his Caltech laboratory on 2 August and published first.[89] This new particle, dubbed the positron, was the first anti-particle ever found. A little later, Blackett and Occhialini discovered that showers of positron-electron pairs were produced by gamma rays — high energy photons — crashing into the upper atmosphere.

Being pipped at the post over the positron did not reduce the fact that 1932 was an extraordinary year at the Cavendish, and Rutherford knew that much more remained to be found. As he remarked in *Nature* in 1933, he and his colleagues were barely scratching the surface: the velocities achieved in the new particle accelerators were not comparable with the velocities of particles in Earth's atmosphere.[90] In short, the new world of high-energy particle accelerators was only just beginning, and by late 1934 Rutherford was wondering whether he could fit an accelerator capable of two million electron volts into the laboratory space.[91]

11 | Elder statesman

THE 1931 NEW YEARS' HONOURS LIST brought fresh accolade for Rutherford: a baronage. He had the option of adopting a new name but elected instead to become 1st Baron Rutherford of Nelson, a recognition of his birthplace and original home. His coat of arms contained New Zealand motifs. It was a clear statement: nearly half a century after leaving Nelson and 40-odd years after leaving New Zealand, now at the very pinnacle of British society, he still drew identity from the place of his childhood. He cabled his mother: 'Now Lord Rutherford, more your honour than mine, Ernest.'[1] It was a singular moment, underscoring just what a colossus Rutherford had become. Many leading British and Imperial scientists were knighted, but few became peers.

Rutherford did not let it go to his head. His brothers continued to call him Ern, as they always had.[2] He remained 'Cousin Ernest' to a relation, Eileen 'Bay' de Renzie, who was living in the Rutherford home at the time.[3] He made no real effort to collect memorabilia and newspaper cuttings. Bay wrote to Rutherford's mother for any New Zealand reports of the honour.[4] Others were more impressed and Rutherford was deluged with congratulations. A list he made to manage his replies runs to more than three pages, including Soddy, Oliphant, Thomson, Broad, Andrade, Bragg and dozens more from his own field.[5] Letters came from far and wide. The Governor-General of New Zealand, Lord Bledisloe, told him that he was 'extremely jealous' of his friend Lord Crawford having the 'privileged experience' of introducing Rutherford to the House of Lords. 'New Zealand continues ... to express its profound pleasure at the well-merited recognition of your outstanding achievements.'[6] 'Dear "Baron",' a letter from an old New Zealand school friend opened, 'please accept my sincere congratulations on another honour.'[7] D'Arcy Thompson told him the recognition joined his other achievements and was 'above the reach of envy.'[8] Rutherford seems to have spent a large part of January 1931 answering all these letters.[9]

For Rutherford the award was overshadowed by personal tragedy. His daughter Eileen had married the physicist Ralph Fowler in 1921. In December 1930 she gave birth to her fourth child, but died suddenly 11 days later, aged 29. The loss of his only child was a heavy blow for Rutherford, and all the more poignant because Mary was far away in New Zealand.[10] Rutherford fielded letters of sympathy from afar, even from old school friends such as Charles Broad, who heard the news via contacts in Christchurch where Mary was staying.[11] The Canadian physicist John McLennan (1867–1935) wrote from Toronto to say he and his wife were 'shocked and deeply pained' at the news.[12] Geiger — who knew all the Rutherfords well — felt 'very severely the heavy loss to your family ... How hard for you that Lady Rutherford was so far away in those days of fear and sorrow, and how hard for Lady Rutherford too.'[13]

By this time Rutherford was very much the elder statesman of physics, a point made clear when he turned 60 and Max Born wrote to him with 'heartiest congratulations', expressing his 'friendship and devotion to the great leader of our science.'[14] His standing had reached the point where he was able to write to the Lord President of the Privy Council, Stanley Baldwin, ask him to open a new laboratory wing, and receive a personal handwritten confirmation.[15] Less rewarding was the fact that as a public figure he was pressed by the media for feature articles on the work he and his associates were doing.[16] The calls on his time mounted up.

During the early 1930s Rutherford joined the Academic Assistance Council to support Jewish scientists, who became refugees from the Nazi regime that came to power in Germany in 1933.[17] He became president, actively working to raise funds from the commercial sector in support of the refugees.[18] The problem of what became known as 'the wandering scholars' continued to grow.[19] Money was difficult to raise, and he did not have the connections or opportunities to help everybody: as he told F.G. Donnan, 'I am continually bombarded by requests to help University refugees'.[20] Rutherford nonetheless did what he could, feeling that the remit went beyond those merely displaced by Europe's growing totalitarianism. That spanned ideological divides. Rutherford never did have much patience with communism, but he went out of his way to help Kapitza, who had returned to the Soviet Union in 1933 purely to visit,[21] only to then be forbidden to leave. After some thought Kapitza restarted his work in Moscow, eventually obtaining equipment with Rutherford's help.[22]

Rutherford's colleagues and students could not imagine a world without him. He was the go-to expert in the field — such as the time he suggested names to US physicist Harold Urey for the stable hydrogen isotope deuterium that Urey discovered in 1933. This isotope was a hydrogen atom with nucleus comprising a proton *and* a neutron. Initial proposals were derived from the Greek 'deuteros', but Rutherford didn't like them. To him 'Deuton' seemed too easy to confuse with 'neutron', while 'Deuterogen' smacked, as Rutherford put it, of 'Diogenes and his tub.' Instead Rutherford suggested 'Deuteron' which was 'a better word in all respects.'[23] This word was adopted to mean the ion form of the isotope. Meanwhile the isotope itself — a hydrogen atom with nucleus consisting of a proton and a neutron — eventually became called deuterium. The more common form of hydrogen with its single proton in the nucleus was sometimes called protium.

WHY RADIOACTIVITY OCCURS

CHADWICK'S DISCOVERY of the neutron, at Rutherford's urging, brought together the final major piece of the atomic puzzle. An atomic nucleus comprised two sorts of particles: protons — which carried a positive charge — and neutrons which, as the name suggests, didn't. Around the nucleus were the electrons, which quantum mechanics had shown existed as standing waves and thus became a 'cloud' around the nucleus.

The stability of this structure relies on charge and force balances between the components. Atoms where the protons, neutrons and electrons are in equilibrium are stable. But some elements are not stable — for example, radium has 88 protons, 138 neutrons and 88 electrons. The ratio of protons to neutrons is part of the mixture: lighter elements require a 1:1 relationship, while heavier ones from calcium onwards on the periodic table are usually in 1:1.5 balance. It is not precise: some elements can be stable with differing numbers of neutrons, creating different isotopes.

This process also explains the forms of radiation that Rutherford discovered. With alpha radiation an unstable nucleus emits two protons and two neutrons together — a helium-4 nucleus. Beta radiation is more complex than was known in Rutherford's day. The fact that it is an electron was well known by the time Rutherford began investigating radioactivity in earnest, but its origin was not. Today we know that it is produced by particle transformation in the nucleus as a result of the weak force, which was proposed in 1933 by the Italian-American physicist Enrico Fermi (1901–1954) but not fully explored until the 1950s and 60s. Gamma radiation is composed of very high-energy photons emitted by a nucleus as a result of a decay event. All are symptomatic of the process of radioactive decay, whereby entropy drives unstable atoms to decay into a stable form and emit particles or electromagnetic radiation.

This shows, yet again, how fundamental Rutherford's discovery of the nucleus was to the whole of modern physics and chemistry.

Rutherford was often approached by colleagues for comment, as in April 1934 when Gamow — then working with Bohr in Copenhagen — sent through a draft paper on the possibility of 'negative protons'.[24] Other requests were less directly in his field but make clear he was viewed as the font of wisdom about all matters science. As one example, he was asked in 1934 about proposals to test heavy water — water made with deuterium — on mice. He immediately visualised how such an experiment might work. 'I should have thought that initially one would want to observe the effect of heavy water in different concentrations ... on a good many animals and plenty of controls.'[25] He continued to push research at the Cavendish, sending Oliphant and Cockcroft to the Philips labs in the Netherlands to investigate their equipment, which included a million-volt accelerator.[26]

By this time Rutherford was getting on, now rather portly and distinctly white-haired. He declared that he intended to retire at 70, though it is difficult to imagine he would have let physics go. The pressures on his time at the Cavendish remained intense. In late 1934 Rutherford received a heartfelt message from his mother, via Ernest Walker: Martha Rutherford had asked him to 'try and influence you to come out to New Zealand to see her again before she crosses the river'. It was 'unusual for her to seek help in her private affairs' but 'she so longs to see you that in the present issue she made an exception to the rule ... She said that "To the world Ernest is a great scientist but to me he is my son."'[27] Martha even argued that if time was short, her son could always fly to Australia.[28] It was a cogent, desperate plea from a mother to her son. But Rutherford did not make the journey, and his mother died in July 1935, aged 92.[29]

Accolades and awards continued to pour Rutherford's way. He was elected president of the British Association for the Advancement of Science. When Pope Pius XI established the Pontifical Academy of Sciences in 1936, Rutherford was elected a founding member and issued a membership card in a small gold-engraved wallet.[30] By this time the world was entering a new era of instability. Fascism was on the rise in Europe. Britain began re-arming, and in early 1937

Rutherford was asked to join a proposed research council, reporting to the Committee for Imperial Defence on scientific matters — something now well recognised as vital to military technology.[31]

By this time Rutherford and Mary had a small holiday home, a little cottage at Chute in Wiltshire, on a property he was delighted to find was called 'New Zealand Farm', though it was run by 'an Englishman who has been in Canada.'[32] They got away to it whenever they could. In September 1937 they spent Rutherford's 66th birthday there. He was back in Cambridge by 5 September, when he wrote to Geiger to say he felt in 'good form' and was busy 'tree cutting, motoring etc.' He was preparing to attend the Indian Science Congress being held early in 1938, with the British Association for the Advancement of Science. He expected that he and Mary would 'enjoy this first visit to India.'[33] As president-elect of the Association he had to prepare a speech in which — yet again — he laid out his hopes for future research, this time using new high-powered equipment to study matter transmutation. From this Rutherford expected to learn more about the structure of 'different forms of atomic nuclei and the origin of the elements.'[34]

Rutherford was at his cottage on 1 October, where he cleared blackberry,[35] but back again in Cambridge by 14 October, when according to his brother's later account, he apparently 'got up feeling a bit mouldy.'[36] That day he probably received a letter Oliphant had written three days earlier after taking up a new post in the University of Birmingham, lamenting how he missed Cambridge.[37] Rutherford ate nothing during the day and called his colleague Norman Feather to his house. Mary served tea. Rutherford asked Feather to take his next lecture for him, did not eat the cake Mary provided, and handed Feather his lecture notes. Then he did something unusual. Feather, years later, vividly recalled the moment. As they parted Rutherford shook his hand, hesitantly and awkwardly. Feather was astonished: Rutherford never shook hands normally.[38]

Rutherford apparently put his illness 'down to some roast pork they had for dinner the night before,'[39] had a little soup for dinner, but was then sick. He went to bed around 10 p.m. and was ill overnight. Around 7.15 a.m. next day, 15 October, Mary called his doctor, Henry ('Harry') E. Nourse, who suspected a small hernia had become strangulated and was obstructing Rutherford's intestines. Nourse sent Rutherford to the Evelyn Nursing Home, a private hospital in Cambridge, and called John Ryle, the Regius Professor of Physic [medicine], who concurred. However, surgery had to wait on the arrival of Sir Thomas Dunhill, who was called up from London and did not see Rutherford until 6.45 p.m. Rutherford was concerned that any surgery might prevent him travelling to India — however, it had to be done. Dunhill 'was satisfied the operation was in time.'[40] But Rutherford never recovered.[41] During the afternoon of 19 October Mary wrote to Oliphant that her husband was being a 'wonderful patient' and a 'thread of hope' remained.[42] But it was only a thread, and Rutherford died that evening, peacefully, about an hour after declaring that Nelson College was to receive £100 from his estate.[43]

'Papa' was dead. The news stunned the physics community. Those who had worked with him for so many years, who had become accustomed to his wisdom and guidance, felt it not as the loss of a friend and colleague, but as the death of a parent. Bohr, Cockcroft and Oliphant were at a congress being held in Bologna to mark the 200th anniversary of the birth of Luigi Galvani.[44] A telegram arrived from Cambridge. Bohr, barely able to control his tears, stood before the conference and with quavering voice gave the news to the shocked gathering.[45] Tributes rolled in thick and fast. J.J. Thomson declared Rutherford's death one of the greatest losses British science had ever experienced.[46] The Director of the Science Museum, Ernest Mackintosh (1880–1957) wrote to Cockcroft extending 'sincere condolences' on behalf of his museum. 'His pre-eminence was such that any eulogy of mine would almost be presumptuous, and his sudden passing is a terrible bereavement to this country and to the whole world of science.'[47] The architect Charles Holden (1875–1960) told Cockcroft that 'the charming intimacy of his manner, so unself-conscious, was one of the most lovely attributes of this great man.'[48]

At the request of the Royal Society Rutherford was buried in Westminster Abbey, Britain's national cathedral. The funeral service on 25 October was solemn, poignant and wrapped in all the ceremony of the British establishment. It began with music: Bairstow's Elegy, chorale preludes by J.S. Bach, and Charles Parry's Elegy among others. Mary Rutherford, beset with grief, was supported by Bohr and Phyllida Cook.[49] There were prayers, a lesson, and hymns. Rutherford's ashes were carried by ten pallbearers, including the High Commissioner of New Zealand, the vice chancellor of Cambridge University, the president of the Royal Society, and the president of the Royal College of Physicians. He was interred near the ashes of Newton and of Lord Kelvin. There was a blessing, and the ceremony ended to the haunting tones of Basil Harwood's 'Requiem Aeternum.'[50] There was a separate memorial service in the chapel at Trinity College. In New Zealand the Minister of Agriculture gave a tribute in the House: 'Lord Rutherford was one of the greatest, if not the greatest, of New Zealand's sons, and his passing leaves the world much poorer.' Students of Nelson College attended a special memorial service in Nelson Cathedral.[51] And in Moscow, Peter Kapitza led a gathering of senior scientists to remember their fallen friend and mentor.[52]

Mary went to the cottage at Chute a week after Rutherford's funeral and found it 'more desperately sad' than their house in Cambridge.[53] Finally she accepted an offer of a holiday in South Africa. Mary Rutherford was not alone. For all those who knew Rutherford, who had been inspired by his energies and who looked on him as the father figure of science, life was never the same again.

12 | Ernest Rutherford and the birth of modern physics

I expressed to him my admiration at the breadth and depth of his knowledge in almost every branch of physics and chemistry. 'Ah! Mr Marsden,' he replied, 'When I want to know about a subject I put myself down to lecture on it. Then I have to study it.'

— Ernest Marsden[1]

ERNEST RUTHERFORD'S WORK SPANNED the four decades or so that produced the basic frameworks of modern physics. It was a new vision of the universe that overturned classical assumptions and was built around the concepts of particles and nuclear atoms that Rutherford did so much to pioneer. In subsequent decades the operating systems for Rutherford's particles — Bohr, Dirac and Heisenberg's quantum mechanics at micro-level, Einstein's relativity at macro-scale — were further explored and proven. From this emerged a vaulting concept of an expanding universe that erupted out of nothing in a 'big bang', of spinning black holes that twisted the very fabric of space-time around on itself, of particles that seemed entangled with each other irrespective of separation, and of particles that appeared and disappeared so quickly that they might be said to have never existed at all.

Had Rutherford survived to the Second World War he would doubtless have become involved in the efforts to develop better radar and other technologies. In the event the legacy of his work contributed directly to the Manhattan Project, the Allied push to weaponise the enormous energies Rutherford and his colleagues had discovered within atoms. Part of this task involved harnessing arrays of cyclotrons to transmute one uranium isotope into another. This process, dubbed 'enrichment', applied the principles Rutherford had defined, this time in order to fuel a bomb designed to split atoms in a chain reaction, using the neutron he had predicted and which Chadwick had found. What Rutherford might have thought of this is clear from his remarks about such weapons during the First World War, and we can be certain he would have viewed the Manhattan Project with horror.

A few days after Ernest Rutherford died, his brother James wrote to Marsden, who by this time was head of the Department of Scientific and Industrial Research, proposing a biography.[2] Marsden had been interviewing Rutherford's New Zealand relatives in a general effort to collate material for a possible future volume and wrote to Mary urging that such a book should be written 'by as competent a person as possible', worrying that he did not have the competence himself, 'though nothing would be more to my liking.'[3] Mary and Rutherford's other executor, Ralph Fowler, liked the idea but preferred to keep it local to Britain, asking Arthur Eve if he might prepare an official account. They obtained a swift buy-in from Cambridge University Press. Meanwhile, Marsden began collecting material from Rutherford's New Zealand family.[4] Within a few days he felt that even if somebody in Britain wrote a biography, the material he was gathering was not available there and would support a separate book focusing on his New Zealand life.[5]

Mary eventually wrote to Marsden asking him not to write a book — there didn't seem to be space for two volumes. Eve completed the official volume in 1939. Marsden continued to toy with writing one himself, even getting involved in a series of arrangements with Faber & Faber in the late 1950s.[6] He left the material he had collected among his papers. These were eventually deposited in the National Library of New Zealand, but with a few notable exceptions this material was either overlooked or not consulted by Rutherford's later biographers. The folios contain a mass of material ranging from reminiscences of Rutherford's childhood penned four and five decades after events, to character studies by contemporaries.[7]

This material is an important resource: from this a character emerges, clearly somewhat rose-tinted but with a consistency and a human quality that gives credibility to the picture. And it offers some explanations as to the nature of the man. Rutherford was conservative, often literal-minded — a characteristic that also framed his science and pushed him into the hands-on practical world of experiment. He was impatient, a characteristic that intruded particularly when he had questions to answer. But he was also gregarious. He liked being around people, drawing inspiration and energy from those around him. That stood in some contrast to the usual image of physicists as introverts drained by human contact and who actively sought time alone. This joy of social contact with his colleagues was almost certainly a key part of how and why Rutherford became a father figure to his students and colleagues.

Rutherford does not seem to have been particularly neurodiverse. The idea of neurodiversity as a key to genius has become a popular trope of the twenty-first century. In some respects Rutherford clearly was not: he fitted well enough into school, a sharp contrast to figures of his day known to be neurodiverse, such as Churchill or Einstein. Nonetheless, Rutherford showed some classic aspects of 'thinking different'. At high school he focused intensely on what was before him — typically a book he was reading — to the point where other students discovered they could startle him by banging him on the head with a book before, as Charles Broad put it, 'bolting for our lives'.[8] Rutherford's ability to 'zone out' has since been associated with attention deficit disorder, part of the neurodiversity spectrum. He also had excellent mathematics skills, and mathematics is the language of the abstract shapes and patterns by which some human minds think. Certainly it seems clear that Rutherford thought in such shapes. His difficulty was translating the symbolisms of his mind into the linear sequence of words required to express them. This came out particularly when, as a university student, he briefly taught high school maths. A former pupil, G. Gillespie, found him 'entirely hopeless as a schoolmaster'.[9]

Similarly, as Rutherford's fellow university student William Marris observed, Rutherford was 'nervy and excited' during examinations, due — Marris suspected — to 'bookwork'.[10] And while he had no obvious problem reading and writing, he missed detail. That was a lifetime issue: as one editor noted, Rutherford was a 'very indifferent proofreader' of his own papers.[11] He also showed signs of being somewhere on the dyslexic spectrum, with effects on word retrieval — notably his occasional mix-up of words when speaking. And he had little interest in rote learning: decades later, he told his own students that he was teaching them how to think, not to remember, nor did he intend to spoonfeed them.[12]

Rutherford's problem articulating his thoughts led to lasting misconceptions about his capabilities. To listeners he presented as ordinary: there was, as Coleridge Farr declared later, 'nothing of the typical "highbrow" about him.'[13] He was easily dismissed as a plodder, competent but uninspired, getting results by hard work. His difficulty expressing thoughts on the spot likely contributed to his early botched job interviews and the way he struggled with exams in his youth.

All this masked Rutherford's extraordinary genius, his ability to conceive of a problem and to seize upon a piece of evidence that others did not — always the key to solving primary problems in physics. As became evident during his research on radioactivity, that process usually began for Rutherford with some shrewd ideas as to why the phenomena he saw were behaving as they did, from which he then devised equipment that could analyse it in unique ways, giving him numeric data from which he could develop a conclusion. Along the way he continued to think through the problem, often coming up with a soaring vision that then guided the way forward. Frequently, as we have seen through this book, he revealed glimpses of the big picture developing in his mind, but he was too careful and thorough a scientist to render that in his papers, instead constraining himself to what he could show to be true for now.

This approach was not obvious. Few outside Rutherford's closest friends and colleagues realised that, for all the vagueness with which he sometimes spoke, he had a clear general vision in mind. This vision — combined with his outstanding ability to create measuring equipment with which to put numbers to his ideas — set his work apart from that of his peers. Rutherford's weakness, if it could be called that, was his realism. To him the universe was one of real particles and real forces, a universe that was relentlessly determinist and thus relentlessly measurable. His focus on experimentation and hard data, on the literal realities that he could see from his instruments, increasingly stood apart from the more abstruse theoretical paths taken by quantum mechanics during the 1920s, with their concepts of wave-function collapse and uncertainties.

In short, Rutherford was a man of extraordinary intellect, but one not readily recognised in a Western environment where intellect was usually identified with quick wit and ability to recall obscure data. Rutherford did not fit that expectation, masking the genius of his analysis, ideas and methods. His unique abilities, particularly once backed by the resources of major university departments, set him up for a remarkable series of discoveries.

One point often missed historically is that Rutherford's primary work was also a consistent path of investigation with coherent purpose in which seemingly separate achievements, such as his discovery of radioactive decay and his later discovery of the atomic nucleus, were actually aspects of the same line of investigation. Although often sliced and diced by history, thanks in part to the way that physics itself is compartmentalised, Rutherford's work from his earlier days was a continuity, a relentless push down a specific path in which each discovery fed into the next. He began at Canterbury with work in electrodynamics, study that gave him a sound technical basis for the instrumentation he then used to support his investigations into ions and radioactive ores at Cambridge. That in turn fed into his remarkable exploration of radioactivity at McGill. Once he had determined that radioactivity was due to atoms spontaneously breaking down he wanted to know how and why, which took him directly to the centre of the atom and his discovery of the atomic nucleus. This of itself was only part of the picture, so he then turned to investigating the nature of that nucleus in order to find the full explanation for what he was observing. His landmark achievements along the way — identifying radioactivity as atomic breakdown, discovering the atomic nucleus, and then splitting the atom — were waypoints in a consistent journey of exploration across a single broad theme. He continued that work at the Cavendish in the 1920s, this time as director and conductor of a research establishment. Even what seemed to be diversions, as when Rutherford looked into the age of the Earth, were in reality part of the broader mix, in this case a direct outcome of his concept of half-life.

The realities that Rutherford discovered about the subatomic world then fed into the theoretical physics being developed by Einstein, Bohr, Heisenberg and others, eventually interrelating with the whole of the new physics. As we saw in chapter 1, the nearest theory today to a 'theory of everything' remains the Standard Model of particles, a model directly founded in Rutherford's work and in the directions he took, including the way he swung the focus of the Cavendish into that area during the 1920s and 30s.

The picture that emerges is clear. Rutherford is remembered as a pioneering nuclear physicist, but historically he was more than that. His discovery of the atomic nucleus was a key factor in a ground-shift behind the basis on which chemistry was understood. His investigation of the nature of that nucleus established directions that led to the discovery of new fundamental forces and the main principles of particle physics. This field stands at the core of our understanding of physics today. Rutherford did not do all this alone, but his contribution to the work of those around him is also clear. Although he often refused to take due credit, his energy, curiosity, sense of enquiry and relentless pursuit of data stood behind the work of many of his contemporaries. As head of university departments and, later, the Cavendish, Rutherford was also the primary organiser of the direction and nature of wider research, shaping that research while fuelling his staff with an enthusiasm and drive that meant his influence transcended his own hands-on work. In short, Rutherford was a colossus in the field by any measure, one of the key players who set modern physics on its way.

Since then, many of the answers to questions posed by Rutherford and his colleagues have been found, discoveries that show how the foundations he established in nuclear physics remain central to understanding the workings of the universe. That journey is another story — but some of the initial steps along that path were taken by Rutherford's students and colleagues after his death. And, as Marsden put it later, as these scientists wondered how to understand the mysteries of the universe being laid out before them, how to comprehend their observations and find new ways of exploring the unknown they would often ask themselves: 'What would Rutherford do?'[14]

Notes

Abbreviations

CUL Cambridge University Library (UK)
ATL Alexander Turnbull Library (NZ)
RP Rutherford Papers
n.d. no date

Introduction

1 ATL MS-Papers-1342-260 Marsden, Ernest (Sir), 1889–1970, papers, 'Addresses and articles on Lord Rutherford', Rutherford Jubilee International Conference, September 1961, extract from conference summary, typescript.
2 *New York Times*, 20 October 1937.
3 ATL MS-Papers-1342-254 Marsden, Ernest (Sir), 1889–1970: Papers, Correspondence between Rutherford and Marsden, Typescript, 'Extracts from Forty Years of Physics by the late Lord Rutherford'. Specifically, Rutherford showed that the charge on the electron correlated with what Planck required it to be for the quantum hypothesis to be valid.
4 While at McGill University, ATL MS-Papers-1342-258 Marsden, Ernest (Sir), 1889–1970: Papers. 'Facts and particulars concerning the early life of Lord Rutherford', MS 'Lord Rutherford 1871–1937'. See also, e.g. CUL RP Add 7653, Rutherford Correspondence M1-54, McLennan to Rutherford, 27 July 1935.
5 Einstein made the remark to Chaim Weizmann, see Chaim Weizmann, *Trial and Error: the autobiography of Chaim Weizmann, Vol. 1*, Jewish Publication Society of America, Philadelphia 1949, p. 118.
6 As spoken by James Jeans (1877–1946) to the 1938 Indian Science Congress, see, e.g. Gilbert J. Fowler, 'The Silver Jubilee of the Indian Science Congress', *Current Science*, Vol. 6, No. 7, January 1938, p. 320.
7 S. Devons, FRS, 'Rutherford and the science of his day', *Notes and Records of the Royal Society of London*, Vol. 45, No. 2, (1991), p. 221.
8 ATL MS-Papers-1342-256 Marsden, Ernest (Sir), 1889–1970: Papers, Correspondence re-Lord Rutherford, MS note T. Alty, n.d.
9 Marsden established the Physics Department at Victoria University.
10 In 1991, while working for the Reserve Bank of New Zealand.
11 Supplied by Dr John Campbell of the University of Canterbury. I traced it to the Manchester studio as part of due diligence relative to copyrights. A carte-de-visite was a small photographic portrait popularised from the 1850s, typically produced in multiple copies for the owner to send to friends, family and others.
12 CUL RP Add 7653, Misc. files and notes PA308, Lord Rutherford, 'Early Days in Cambridge', typescript.

1 The summer of '96

1 See www.phy.cam.ac.uk/history/old, accessed 21 May 2024.
2 Described by Rutherford in a letter to his mother, 2 December 1896, in A.S. Eve, *Rutherford, being the life and letters of the Rt Hon Lord Rutherford, OM*, Cambridge University Press, Cambridge, 1939, p. 41.
3 Rutherford to his mother, July 1896, in Eve, p. 37.
4 CUL RP Add 7653, Correspondence G1-80, Gardiner to Rutherford, 14 January 1932. Gardiner had been Crookes' assistant when Crookes developed the tubes and was responding to a request by Rutherford for information.
5 blog.sciencemuseum.org.uk/jj-thomsons-cathode-ray-tube/, accessed 21 May 2024.
6 Thomson, J.J. 'Cathode Rays', *The London, Edinburgh and Dublin Philosophical Magazine and Journal of Science*, 5th Series, Vol. 44, 1897.
7 The effect was later named 'cathodoluminescence' and is an outcome of quantum mechanics.
8 Röntgen's name was selected in 1928 for a unit by which to measure exposure to X-rays. The equivalent SI term, adopted in 1974, is 'coulomb per kilogram'.
9 Rutherford to his mother, July 1896, in Eve, p. 37.
10 J.J. Thomson and E. Rutherford, 'On the passage of electricity through gases exposed to Röntgen rays', *The London, Edinburgh and Dublin Philosophical Magazine and Journal of Science*, 5th Series, Vol. 42, July–Dec 1896, pp. 398–99.
11 Rutherford to Mary, 24 April 1896, quoted in Eve, p. 34.
12 He was thanked for it in their joint paper: J.J. Thomson and E. Rutherford, 'On the passage of electricity through gases exposed to Röntgen rays', *The London, Edinburgh and Dublin Philosophical Magazine and Journal of Science*, 5th Series, Vol. 42, July–Dec 1896, p. 407.
13 'Ionising' radiation dislodges electrons from whatever atoms it strikes. This causes cell damage. See www.ncbi.nlm.nih.gov/pmc/articles/PMC3520298/, accessed 23 May 2024.
14 Summarised in Livia Gershon, 'The X-ray craze of 1896', daily.jstor.org/the-x-ray-craze-of-1896/, accessed 22 May 2024.
15 H.G. Wells, *The War of the Worlds*, Harper & Brothers, New York, 1898, p. 181. This was written in 1895–96, serialised in 1897 in *Pearson's Magazine*, then published as a book in 1898.
16 sciencenotes.org/when-were-the-elements-discovered-timeline-and-periodic-table/, accessed 6 April 2025.
17 ATL MS-Papers-1342-260, Marsden, Ernest (Sir), 1889–1970, Papers, 'Addresses and articles on Lord Rutherford', Ernest Rutherford, 'Lecture on the structure of the atom', typescript, n.d.
18 For example, J.J. Thomson, E. Rutherford, 'On the passage of electricity through gases exposed to Röntgen rays', *The London, Edinburgh and Dublin Philosophical Magazine and Journal of Science*, Vol. 42, Issue 258, 1896.
19 In the modern Standard Model, electrons are classed as 'Fermions', particles whose spin is defined by half-integers.
20 The charge values found in the 1890s have since been revised. Electrons specifically carry a charge of 1.602176634 × 10–19 coulomb. All subatomic particles have either this charge or a whole-number multiple of it.
21 For FDA report, see www.fda.gov/radiation-emitting-products/resources-you-radiation-emitting-products/television-radiation, accessed 22 May 2024.

22 *The Big Bang Theory*, CBS 2007–2019.
23 Einstein's letter is at www.osti.gov/opennet/manhattan-project-history/Resources/einstein_letter_photograph.htm#1, accessed 22 May 2024.

2 The origins of Rutherford's science

1 From an address Rutherford gave to the Royal Academy of Arts, quoted in Eve, p. 353.
2 CUL RP Add 7653, Misc. files and notes, PA312, Hahn & Meitner, 'Lord Rutherford's Sixtieth Birthday', typescript.
3 As summarised in Joeri Witteveen, 'Objectivity, Historicity, Taxonomy', *Erkenntnis* Vol. 83, pp. 445–63, 2018, via doi.org/10.1007/s10670-017-9897-z, accessed 5 July 2024.
4 Key contribution to the debate was Steven Shapin, *The Scientific Revolution*, University of Chicago Press, Chicago, 1996; also Thomas E. Kuhn, *The Structure of Scientific Revolutions*, University of Chicago Press, Chicago, 1962, 3rd edition, 1996.
5 See, e.g. Helen Longino, *The Fate of Knowledge*, Princeton University Press, Princeton, 2002; for analysis of continued discussion see Mansoor Niaz, 'Evolving Nature of Objectivity in the History of Science and its Implications for Science Education', 2018; 46: 37–77, published online 29 October 2017, DOI: 10.1007/978-3-319-67726-2_3.
6 Philippe Stamenkovic, 'Facts and objectivity in science', *Interdisciplinary Science Reviews*, 48(2), pp. 277–98. doi.org/10.1080/03080188.2022.2150807.
7 The 'Eureka' story only emerged later: see David Biello, 'Fact or fiction? Archimedes coined the term "Eureka!" in the bath', *Scientific American*, 8 December 2006, www.scientificamerican.com/article/fact-or-fiction-archimede/, accessed 24 June 2024.
8 For details see Philip Kitcher, 'Fluxions, Limits and Infinite Littleness: a study of Newton's presentation of the calculus', *Isis*, Vol. 64, No. 1, March 1973.
9 *Oxford English Dictionary*, 1st Edition.
10 Quoted in Eve, p. 353.
11 Usefully described by Richard Reeves, *A Force of Nature: the frontier genius of Ernest Rutherford*, Norton, New York, 2008, pp. 13–14.
12 Malcolm Longair, 'Rutherford and the Cavendish Laboratory', *Journal of the Royal Society of New Zealand*, 51(3–4), pp. 444–66, at doi.org/10.1080/03036758.2021.1885452.
13 See chapter 8 for details.
14 L. Barham, Duller, G.A.T., Candy, I. et al. 'Evidence for the earliest structural use of wood at least 476,000 years ago'. *Nature*, 622, 107–11, 2023. doi.org/10.1038/s41586-023-06557-9, accessed 1 July 2024.
15 laria Degano, Sylvain Soriano, Paola Villa, Luca Pollarolo, Jeannette J. Lucejko, Zenobia Jacobs, Katerina Douka, Silvana Vitagliano, Carlo Tozzi, 'Hafting of Middle Paleolithic tools in Latium (central Italy): New data from Fossellone and Sant'Agostino caves', doi.org/10.1371/journal.pone.0213473, 20 June 2019; also Kozowyk, P.R.B., Baron, L.I. & Langejans, G.H.J., 'Identifying Palaeolithic birch tar production techniques: challenges from an experimental biomolecular approach', *Sci Rep* 13, 14727, 2023; doi.org/10.1038/s41598-023-41898-5, both accessed 12 April 2024.
16 Patrick Schmidt, Radu Iovita, Armelle Charrié-Duhaut, Gunther Möller, Abay Namen, Ewa Dutkiewicz, 'Ochre-based compound adhesives at the Mousterian type-site document: complex cognition and high investment', *Science Advances*, 21 February 2024, Vol. 10, Issue 8, DOI: 10.1126/sciadv.adl082, accessed 12 April 2024.
17 C. Tuniz, F. Bernadini, I. Turk, L. Dimkaroski, L. Mancini, D. Dreossi, 'Did Neanderthals play music? X-ray computed micro-tomography of the Divje babe "flute"', doi.org/10.1111/j.1475-4754.2011.00630.x, 28 August 2011; also Marquet J-C, Freiesleben, T.H., Thomsen, K.J., Murray, A.S., Calligaro, M., Macaire, J-J., et al., 'The earliest unambiguous Neanderthal engravings on cave walls: La Roche-Cotard, Loire Valley, France', 2023, *PLoS ONE* 18(6): e0286568; doi.org/10.1371/journal.pone.0286568, accessed 12 April 2024. These discoveries have been subject to due debate.
18 Samantha Brown, et al., 'The earliest Denisovans and their cultural adaptation', *Nature*, 25 November 2021, www.nature.com/articles/s41559-021-01581-2, accessed 26 June 2024.
19 Schmidt, et al., accessed 12 April 2024.
20 Michelle C. Langley and Thomas Suddendorf, 'Archaeological evidence for thinking about possibilities in hominin evolution', doi.org/10.1098/rstb.2021.0350, accessed 10 May 2024; also Jean-Jacques Hublin, Simon Neubauer, and Philipp Gunz, 'Brain ontogeny and life history in Pleistocene hominins ', *Philosophical Transactions of the Royal Society, London*, B. Biol. Sci, 5 March 2015; 370(1663): 20140062 at www.ncbi.nlm.nih.gov/pmc/articles/PMC4305163/, accessed 10 May 2024.
21 ATL MS-Papers-1342-260, Marsden, Ernest (Sir), 1889–1970, Papers, Addresses and articles on Lord Rutherford, Ernest Rutherford, 'Lecture on the structure of the atom', typescript, n.d.
22 General thesis but see, e.g. Kuhn, pp. 43–51, 77–91.
23 Francis Bacon, 'Of Studies', *The Essays Or Counsels, Civil And Moral, Of Francis Ld Verulam Viscount St. Albans*, via Project Gutenberg, www.gutenberg.org/files/575/575-h/575-h.htm, accessed 2 July 2024.
24 Jayant Shah, 'Accuracy of Ptolemy's Almagest in predicting solar eclipses', *Annals of Mathematical Science and Applications*, Vol. 3, No. 1, 2018, pp. 7-29, esp. p. 22.
25 Kuhn, pp. 82–84.
26 royalsociety.org/about-us/who-we-are/history/, accessed 3 July 2024.
27 See Halley to Newton, 22 May 1686, makingscience.royalsociety.org/items/el_h3_44/letter-from-edmond-halley-to-isaac-newton?page=1.
28 For example, Robert Boyle (1627–91), widely regarded as one of the founders of modern chemistry.
29 Usefully summarised in John Ralston Saul, *Voltaire's Bastards: the dictatorship of reason in the west*, Penguin, London, 1993, pp. 37–76.
30 For mathematical details see physics.stackexchange.com/questions/29363/how-did-halley-calculate-the-distance-to-the-sun-by-measuring-the-transit-of-ven, accessed 11 September 2024.
31 For summary see 'Scientists debate planet definition and agree to disagree', www.psi.edu/blog/scientists-debate-planet-definition-and-agree-to-disagree/, accessed 7 July 2024.
32 See, e.g. Anthony S. Travis, 'The accidental discovery of mauve', *Victorian Review*, Vol. 40, No. 2, Johns Hopkins University Press, 2014.

33 The original novel never used the phrase.
34 See nzas.co.nz/, accessed 26 December 2024.

3 Young Rutherford

1 Registration number 1871/310 see www.bdmhistoricalrecords.dia.govt.nz, accessed 10 February 2025.
2 Noted in Eve, p. 373.
3 Ibid, p. 364.
4 E.N. da C. Andrade, *Rutherford and the Nature of the Atom*, Anchor Books, New York, 1964, p. 16.
5 ATL MS-Papers-1242-258 Marsden, Ernest (Sir), 1889–1979: Papers, Facts and particulars concerning the early life of Lord Rutherford, various papers and typescripts.
6 ATL MS-Papers-1342-262 Marsden, Ernest (Sir), 1889–1970: Papers, Addresses and articles written by E Marsden re Lord Rutherford, MS page.
7 ATL MS-Papers-1242-258 Marsden, Ernest (Sir), 1889–1970: Papers, Facts and particulars concerning the early life of Lord Rutherford, untitled typescript.
8 ATL MS-Papers-1342-263 Marsden, Ernest (Sir), 1889–1970: Papers, Addresses and Articles by E. Marsden re-Lord Rutherford, untitled typescript.
9 See, e.g. E. Marsden, 'Rutherford — His Life and work, 1871–1937', Rutherford Memorial Lecture, 1954, *Proceedings of the Royal Society of London*, Series A. Mathematical and Physical Sciences, Vol. 226, Issue 1166, Nov 1954, pp. 284–86.
10 Eve, pp. 1–12.
11 See, e.g., Matthew Wright, *Old South*, Penguin, Auckland, 2009; Matthew Wright, *The Bateman Illustrated History of New Zealand*, 3rd edition, Auckland, 2024.
12 See, e.g., James L. Marshall and Virginia R. Marshall, 'Rutherford and radon', *The Hexagon*, Summer 2010; J.L. Heilbron, *Ernest Rutherford and the explosion of atoms*, Oxford University Press, Oxford, 2003, pp. 10–16. This view is also present in primary material, see e.g. ATL MS-Papers-1242-258 Marsden, Ernest (Sir), 1889–1970: Papers, Facts and particulars concerning the early life of Lord Rutherford, untitled typescript.
13 See, e.g. spark.iop.org/rutherford, accessed 23 August 2024.
14 Lawrence Badash, 'Influence of New Zealand on Rutherford's scientific development', in Nathan Reingold and Marc Rothenberg (eds), *Scientific Colonialism: a cross-cultural comparison*, Smithsonian Institution Press, Washington, 1987, pp. 379–90.
15 See, e.g. Wright, *Old South* pp. 244–56, *Illustrated History of New Zealand*, pp. 193–209, Matthew Wright; *The History of Hawke's Bay*, Intruder Books, Wellington, 3rd edition, 2024, pp. 91–120.
16 See, e.g. Matthew Wright, *Guns and Utu*, Penguin, Auckland, 2011, pp. 44–55.
17 Detailed in Matthew Wright, *Waitangi: A Living Treaty*, Bateman Books, Auckland, 2018, Chapters 1–5.
18 Matthew Wright, *Old South: life and times in the nineteenth century mainland*, Penguin, Auckland, 2009, p. 72.
19 Noted in Campbell, p. 1.
20 M.F. Lloyd Prichard, *An Economic History of New Zealand to 1939*, Collins, Auckland, 1970, p. 145.
21 Wright, *Illustrated History of New Zealand*, pp. 199–209.
22 ATL MS-Papers-1242-258 Marsden, Ernest (Sir), 1889–1970: Papers, Facts and particulars concerning the early life of Lord Rutherford, typescript, 'Particulars supplied by Mr Jas. Rutherford'.
23 Ibid.
24 Ibid. Ira D. Sankey and Dwight Moody were US religious revivalists who published collections of hymns.
25 Ibid, typescript, 'Extract from the Proceedings of the Physical Society', Vol. 50, May 1938, No. 279.
26 *Nelson Evening Mail*, 23 January 1888.
27 ATL MS-Papers-1342-259, Marsden, Ernest (Sir), 1889–1970: papers, typescript, 'Particulars supplied by Mr Jas. Rutherford'.
28 Ibid, 'Some odds and ends of Lord Rutherford's schoolboy days', anon but likely Arthur Rutherford.
29 Ibid, typescript, 'Particulars supplied by Mr Jas. Rutherford'.
30 Ibid, typescript, 'The late Lord Rutherford'.
31 *Marlborough Press*, 14 April 1882.
32 Campbell, p. 20.
33 ATL MS-Papers-1342-259, 'Marsden, Ernest (Sir), 1889–1970: papers', typescript, 'Particulars supplied by Mr Jas. Rutherford'.
34 Ibid.
35 Ibid.
36 Noted in Eve, pp. 33, 42.
37 M.F. Lloyd-Prichard, p. 149.
38 *Marlborough Express*, 24 April 1882.
39 Reserve Bank Inflation Calculator.
40 ATL MS-Papers-10930-2, 'Rutherford family: papers', George Rutherford diaries, 24 January 1885.
41 Geoffrey Brooke, 'Real Wages in New Zealand: 1840–1914', www.nzae.org.nz/wp-content/uploads/2011/08/Real_Wages_in_New_Zealand_1840-1914.pdf accessed 17 September 2023, p. 23.
42 Ibid.
43 *Nelson Evening Mail*, 14 August 1884.
44 Noted in Campbell, pp. 31–32.
45 ATL MS-Papers-1242-258 Marsden, Ernest (Sir), 1889–1970: Papers, 'Facts and particulars concerning the early life of Lord Rutherford', typescript, 'Extract from the Proceedings of the Physical Society', Vol. 50, May 1938, No. 279.
46 *Colonist*, 8 September 1883.
47 *Colonist*, 11 January 1886.
48 ATL MS-Papers-1242-258 Marsden, Ernest (Sir), 1889–1970: Papers, 'Facts and particulars concerning the early life of Lord Rutherford', typescript 'Particulars supplied by Mr. Jas. Rutherford'.
49 Noted in Campbell, p. 13.
50 Ibid, p. 13.
51 ATL MS-Papers-1242-258 Marsden, Ernest (Sir), 1889–1970: Papers, Facts and particulars concerning the early life of Lord Rutherford, typescript, 'Extract from the Proceedings of the Physical Society', Vol. 50, May 1938, No. 279.
52 Eve, p. 2; Balfour Stewart, *Physics*, D. Appleton & Company, New York, 1878.
53 Stewart, pp. 1–7.
54 Ibid, p. 6.
55 Ibid, pp. 56–57.
56 See, e.g. Campbell, p. 13, Reeves p. 26; see also ATL MS-Papers-1242-258 Marsden, Ernest (Sir), 1889–1970: Papers, 'Facts and particulars concerning the early life of Lord Rutherford', untitled typescript; and www.nzedge.com/legends/ernest-rutherford/, accessed 17 September 2024.

57 Stewart, pp. 59–61.
58 www.havelock.school.nz/home/our-history, accessed 12 September 2024.
59 history.aip.org/exhibits/rutherford/sections/young-rutherford.html, accessed 12 September 2024.
60 Campbell, pp. 24–26.
61 William H. Pickering, 'A Geiger counter study of the cosmic radiation', PhD thesis, California Institute of Technology, Pasadena, California, 1936.
62 Quoted in www.theprow.org.nz/yourstory/ernest-rutherford-early-life/, accessed 12 September 2024.
63 ATL MS-Papers-1242-258 Marsden, Ernest (Sir), 1889–1970: Papers, Facts and particulars concerning the early life of Lord Rutherford, typescript, 'Extract from the Proceedings of the Physical Society', Vol. 50, May 1938, No. 279.
64 Ibid, Broad writing in C.C. Farr, 'Early years in New Zealand', typescript.
65 www.theprow.org.nz/society/early-nelson-college/, accessed 19 September 2024.
66 Numbers given by Rutherford in CUL RP Add 7653 PA305a, Notes on W.S. Littlejohn (1934).
67 Winston Churchill, *My Early Life*, Thornton Butterworth, London, 1930, p.17.
68 ATL MS-Papers-1242-259 Marsden, Ernest (Sir), 1889–1979: Papers, 'Facts and particulars concerning the early life of Lord Rutherford', typescript. See also ATL MS-Papers-1242-258 Marsden, Ernest (Sir), 1889–1979: Papers, 'Facts and particulars concerning the early life of Lord Rutherford', 'Extract from the Proceedings of the Physical Society, Vol. 50, May 1937', C.C. Farr, 'Early years in New Zealand', typescript.
69 *Nelson Evening Mail*, 29 October 1887.
70 *Nelson Evening Mail*, 27 December 1887.
71 The property was still on the market in March, see *Colonist*, 1 March 1888.
72 Campbell, pp. 60–61.
73 ATL MS-Papers-1242-259 Marsden, Ernest (Sir), 1889–1970: Papers, 'Facts and particulars concerning the early life of Lord Rutherford', untitled typescript.
74 CUL RP Add 7653 PA305a, Notes on W.S. Littlejohn (1934).
75 ATL MS-Papers-1242-259 Marsden, Ernest (Sir), 1889–1970: Papers, 'Facts and particulars concerning the early life of Lord Rutherford', untitled typescript.
76 Ibid. See also ATL MS-Papers-1242-258 Marsden, Ernest (Sir), 1889–1979: Papers, 'Facts and particulars concerning the early life of Lord Rutherford', 'Extract from the Proceedings of the Physical Society, Vol. 50, May 1937', C.C. Farr, 'Early years in New Zealand', typescript.
77 ATL MS-Papers-1242-259 Marsden, Ernest (Sir), 1889–1979: Papers, 'Facts and particulars concerning the early life of Lord Rutherford', typescript.
78 Campbell, p. 61.
79 ATL MS-Papers-1242-259 Marsden, Ernest (Sir), 1889–1979: Papers, 'Facts and particulars concerning the early life of Lord Rutherford', typescript. See also ATL MS-Papers-1242-258 Marsden, Ernest (Sir), 1889–1979: Papers, 'Facts and particulars concerning the early life of Lord Rutherford', 'Extract from the Proceedings of the Physical Society, Vol. 50, May 1937', C.C. Farr, 'Early years in New Zealand', typescript.
80 Campbell, p. 66.
81 ATL MS-Papers-1242-259 Marsden, Ernest (Sir), 1889–1979: Papers, 'Facts and particulars concerning the early life of Lord Rutherford', typescript.
82 Noted in Campbell, p. 66.
83 ATL MS-Papers-1242-258 Marsden, Ernest (Sir), 1889–1979: Papers, 'Facts and particulars concerning the early life of Lord Rutherford', typescript, 'Extract from the Proceedings of the Physical Society', Vol. 50, May 1938, No. 279.

4 Electrodynamics at Canterbury

1 www.aps.org/apsnews/2008/07/1820-oersted-electromagnetism, accessed 23 September 2024.
2 James Maxwell, 'On Faraday's Lines of Force', *Transactions of the Cambridge Philosophical Society*, Vol. X, Part I, read 10 Dec 1855 and 11 Feb 1856.
3 J. Clerk Maxwell, FRS, [James Maxwell] 'A dynamical theory of the electromagnetic field', *Philosophical Transactions of the Royal Society of London*, Vol. 155, 1865, pp. 459–512.
4 Ibid, p. 465.
5 Albert Einstein, 'Considerations concerning the fundaments of theoretical physics', *Science*, Vol. 91, No. 2369, May 1940, pp. 487–592.
6 Ibid, p. 464.
7 For further details see www.phy.cam.ac.uk/history/old, accessed 27 September 2024.
8 Noted in Salvo D'Agostino, 'Hertz's Researches on Electromagnetic Waves', *Historical Studies in the Physical Sciences*, Vol. 6, (1975), pp. 263–64.
9 Particularly in G.G. Stokes, 'On the Constitution of the Luminiferous Aether, viewed with reference to the phenomenon of Light', *Philosophical Magazine*, Vol. 29, July–December 1846, pp. 6–10; also G.G. Stokes, 'On the Constitution of the Luminiferous Aether', *Philosophical Magazine*, Vol. 32, May 1848, pp. 343–49.
10 This was calculable as soon as the Earth–Sun distance was known as a result of observing the transit of Venus in 1769, as described in chapter 2. The modern figures are 30.29 km/second at periapsis (closest) and 29.295 km/second at apoapsis (furthest).
11 Noted in www.aps.org/apsnews/2024/05/michelson-speed-light-145-years, accessed 29 September 2024. Michelson's measurements were within 0.005 percent of the current accepted value.
12 Hans J. Haubold, 'Albert A. Michelson's Experimentum Crucis 1881 in Potsdam, Germany', arxiv.org/pdf/2111.12176, accessed 26 September 2024.
13 Albert A. Michelson and Edward W. Morley, 'On the Relative Motion of the Earth and the Luminiferous Ether', *The American Journal of Science*, Vol. XXXIV, No. 203, November 1881, pp. 333–45. A supplement to the paper proposed a means of determining solar movement via interferometry.
14 patents.google.com/patent/US334823, accessed 14 October 2024.
15 Noted in interestingengineering.com/lists/10-of-the-most-important-inventions-of-nikola-tesla, accessed 27 September 2024.
16 See, e.g. Charles Süsskind, 'Observations of Electromagnetic Wave Radiation before Hertz', *Isis*, Vol. 55, No. 1, March 1964, pp. 33–34.
17 Charles Süsskind, 'Hertz and the technological significance of electromagnetic waves', *Isis*, Vol. 53, No. 6, Autumn 1965, p. 342.

18 ATL MS-Papers ATL MS-Papers-1242-258 Marsden, Ernest (Sir), 1889–1970: Papers, 'Facts and particulars concerning the early life of Lord Rutherford', typescript, C.C. Farr, 'Early Years in New Zealand'. Campbell, p. 72, indicated five.
19 Matthew Wright, *Western Front: the New Zealand Division 1916–1918*, Reed, Auckland, 2005, pp. 63–64.
20 ATL MS-Papers-1242-258 Marsden, Ernest (Sir), 1889–1970: Papers, Facts and particulars concerning the early life of Lord Rutherford, typescript, C.C. Farr, 'Early Years in New Zealand', notes by R.M. Laing and S. Page.
21 Matthew Wright, *Old South*, pp. 127–29.
22 ATL MS-Papers-1242-258 Marsden, Ernest (Sir), 1889–1970: Papers, Facts and particulars concerning the early life of Lord Rutherford, typescript, C.C. Farr, 'Early Years in New Zealand'.
23 Ibid, notes by R.M. Laing and S. Page.
24 Ibid.
25 Noted in Campbell, p. 87.
26 Ibid, pp. 84–86.
27 ATL MS-Papers 1242-258 Marsden, Ernest (Sir), 1889–1970: Papers, Facts and particulars concerning the early life of Lord Rutherford, typescript, C.C. Farr, 'Early Years in New Zealand'.
28 The New Zealand Rugby Football Union was formed in 1892, amalgamating prior provincial organisations, largely as a consequence of the widespread popularity of the game. See www.nzrugby.co.nz/about-nzr/our-history, accessed 2 October 2024.
29 Quoted in Eve, p. 10.
30 Ibid, p. 128.
31 Campbell, p. 99.
32 Ibid, pp. 97–98.
33 *Press*, 10 August 1891.
34 *Star*, 10 August 1891.
35 Quoted in Campbell, p. 98.
36 Ibid, p. 116.
37 Ibid, p. 122.
38 E. Rutherford, 'Magnetization of Iron by High-frequency Discharges', *Transactions and Proceedings of the New Zealand Institute*, Vol. XXVII, Wellington, May 1895, p. 497.
39 CUL RP Add 7653, Laboratory Notebooks, NB1–6, NB1.
40 E. Rutherford, 'Magnetization of Iron by High-frequency Discharges', *Transactions and Proceedings of the New Zealand Institute*, Vol. XXVII, Wellington, May 1895, p. 505.
41 CUL RP Add 7653, Laboratory Notebooks, NB1–6, NB1.
42 Per Bank of England Inflation Calculator, comparing 1851 to August 2024, www.bankofengland.co.uk/monetary-policy/inflation/inflation-calculator, accessed 2 October 2024.
43 Campbell, p. 151.
44 See CUL RP Add 7653, Laboratory Notebooks NB1-6, NB3. This notebook contains his 1894 material and carries on to his early work in Cambridge.
45 E. Rutherford, MA, BSc, 1851 Exhibition Science Scholar, 'Magnetic Viscosity', *Transactions and Proceedings of the Royal Society of New Zealand*, Vol. 28, 1895, p. 182.
46 Original sketch plans in CUL RP Add 7653, Laboratory Notebooks NB1-6, NB3.
47 E. Rutherford, MA, BSc, 1851 Exhibition Science Scholar, 'Magnetic Viscosity', *Transactions and Proceedings of the Royal Society of New Zealand*, Vol. 28, 1895, pp. 185–87.
48 Ibid, pp. 183, 187.
49 Specifically, 100.1 megahertz, i.e. 100,100,000 cycles per second.
50 ATL MS-Papers-1342-258 Marsden, Ernest (Sir), 1889–1970: Papers, 'Facts and particulars concerning the early life of Lord Rutherford', R.M. Laing, 'At Canterbury College'.
51 For example, a standard 802.11n or 802.11ac standard wireless router has a range of approximately 25 metres.
52 See, e.g. R.F. Pocock, G.R.M. Harratt, *The Origins of Maritime Radio*, H.M. Stationery Office, London, 1972, p. 1.
53 ATL MS-Papers-1342-258 Marsden, Ernest (Sir), 1889–1970: Papers, 'Facts and particulars concerning the early life of Lord Rutherford', Sir J.J. Thomson, 'Rutherford when a research student at Cambridge', typescript.
54 E. Rutherford, MA, BSc, 1851 Exhibition Science Scholar, 'Magnetic Viscosity', *Transactions and Proceedings of the Royal Society of New Zealand*, Vol. 28, 1895, pp. 182–204.
55 Published as a paper, J.S. Maclaurin, 'On the Action of Potassium-Cyanide Solution upon Gold', *Transactions and Proceedings of the Royal Society of New Zealand*, Vol. 28, 1895.
56 Brian R. Davis, 'Maclaurin, James Scott', *Dictionary of New Zealand Biography*, teara.govt.nz/en/biographies/3m23/maclaurin-james-scott, accessed 3 October 2024.
57 Eve, p. 12; see also Campbell, p. 192.

5 A good keen scientist

1 CUL RP, Misc. files and notes, Add 7653 PA294-299, 'Letter of Application and testimonials in support of the candidature of Ernest Rutherford', E.H. Griffiths; also original typescript in same folio. Roman text per underlining in original.
2 See, e.g. *Marlborough Express*, 3 August 1895.
3 Campbell, p. 192.
4 *Star* (Christchurch), 25 July 1895.
5 Passenger list, *Lyttelton Times*, 1 August 1895.
6 Noted in Eve, p. 13.
7 Ibid, p. 13.
8 CUL RP Add 7653, Correspondence T1-54, Thomson to Rutherford, 24 September 1895.
9 Letter quoted in Eve, p. 15.
10 CUL RP Add 7653, Correspondence T1-54, Thomson to Rutherford, 24 September 1895.
11 Eve, p. 15. Dates calculated by me.
12 For discussion see Matthew Wright, *Illustrated History of New Zealand*, pp. 97–146.
13 In Rutherford's time, New Zealand's settlement by Polynesia had been dated to approximately 1350. Subsequent work has refined this to c.1280–1310. There were no prior settlers.
14 Eve p. 18.
15 ATL MS-Papers-1342-258 Marsden, Ernest (Sir), 1889–1970: Papers, 'Facts and particulars concerning the early life of Lord Rutherford', Sir J.J. Thomson, 'Rutherford when a research student at Cambridge', typescript.
16 Ibid, p. 23.

17 As described in Eve, p. 21.
18 Eve, p. 28.
19 Ibid, pp. 23–24.
20 Ibid, p. 35.
21 Quoted in ibid, p. 39.
22 Max Born, *Einstein's Theory of Relativity*, revised edition, Dover Publications, New York, 1962, pp. 207–208.
23 H. Langevin-Joliot, 'Radium, Marie Curie and Modern Science', *Radiation Research*, No. 150 (Supplement), S3-S8, 1998.
24 www.wordsense.eu/Georgium_Sidus/, accessed 10 April 2025.
25 H. Langevin-Joliot, 'Radium, Marie Curie and Modern Science', *Radiation Research*, No. 150 (Supplement), S3-S8, 1998.
26 Noted in history.aip.org/exhibits/curie/brief/06_quotes/quotes_08.html, accessed 16 October 2024.
27 See, e.g. www.nobelprize.org/prizes/physics/1903/ceremony-speech/, and www.nobelprize.org/prizes/physics/1903/becquerel/lecture/, accessed 25 October 2024.
28 I am grateful to my colleague Lemuel Lyes for his work examining these letters, which required gloves and careful hand-washing. The passage of 120-odd years had diminished the intensity of the radioactivity by only about five percent.
29 www.iflscience.com/marie-curies-body-was-so-radioactive-she-was-buried-in-a-lead-lined-coffin-69080, accessed 21 October 2024.
30 J.J. Thomson, E. Rutherford, 'On the Passage of Electricity through Gases exposed to Röntgen Rays', *The London, Edinburgh and Dublin Philosophical Magazine and Journal of Science*, 5th Series, Vol. 42, July–December 1896, pp. 392–407.
31 Cited in Eve, p. 39.
32 Quoted in ibid.
33 E. Rutherford, 'On the Electrification of gases exposed to Röntgen Rays, and the Absorption of Röntgen Radiation by gases and vapours', *The London, Edinburgh and Dublin Philosophical Magazine and Journal of Science*, 5th Series, Vol. 43, January–June 1897, pp. 241–56.
34 Eve, p. 41.
35 See, e.g. CUL RP Add 7653, Laboratory Notebooks, NB1–6, NB3.
36 Quoted in Eve p. 46.
37 E. Rutherford, 'Uranium Radiation and the Electrical Conduction produced by it', *The London, Edinburgh and Dublin Philosophical Magazine and Journal of Science*, 5th Series, Vol. 47, Jan–June 1899, pp. 109–163, esp. p. 111.
38 Ibid, p. 124.
39 Quoted in Eve, p. 47.
40 www.bankofengland.co.uk/monetary-policy/inflation/inflation-calculator, Bank of England Inflation Calculator, comparing 1897 values with Q3 2024.
41 Quoted in Eve, p. 57.
42 Quoted in ibid, p. 50.
43 Ibid, p. 51.
44 See, e.g. CUL RP Add 7653 Misc. files and notes, PA 294-299, 'Letter of Application and testimonials in support of the candidature of Ernest Rutherford.'
45 Ibid, 28 May 1898.
46 Ibid, E.H. Griffiths.
47 Eve, p. 51.
48 Ibid, p. 55.
49 CUL RP Add 7653 Misc. files and notes, PA294-299, Ball to Rutherford, 20 May 1898. This was an enclosure with his testimonial.

6 Tom Tiddler's ground

1 Rutherford to his mother, 5 January 1902, in Eve, p. 80.
2 Devons, 'Rutherford and the Science of his day', p. 225.
3 J. Hughes, 'Rutherford, Radioactivity and the origins of nuclear physics', *Journal of Physics*, conference series No. 381, 2012, p. 2.
4 See, e.g. Helge Kragh, 'Rutherford, Radioactivity and the atomic nucleus', Centre for Science Studies, Aarhus University, Aarhus, Denmark, n. d., arxiv.org/pdf/1202.0954.
5 Rutherford to Mary, 18 November 1899, in Eve, p. 69.
6 Thomson to Rutherford, 15 February 1901, in Eve, p. 76.
7 Cambridge topped 89.5°F on 17 September, noted on community.netweather.tv/topic/49383-the-remarkable-september-of-1898/, accessed 2 November 2024.
8 First used industrially in the 1830s: orau.org/health-physics-museum/collection/consumer/glass/vaseline-uranium-glass.html, accessed 23 February 2025.
9 Neil Todd, 'Historical and radio-archaeological perspectives on the use of radioactive substances by Ernest Rutherford', University of Manchester, 1 December 2008, p. 12.
10 Rutherford to May, 25 September 1898. Noted in Eve, p. 63. Sometimes given as 'first' class, but the passenger list clearly states 'saloon'.
11 Library and Archives Canada, Passenger Lists 1865–1922, Passenger lists of the YORKSHIRE arriving in Quebec, Que. and Montreal, Que. on 1898-09-18.
12 Ibid.
13 Rutherford to May, 25 September 1898, noted in Eve p. 63.
14 Campbell, p. 249.
15 *New York Times*, 20 September 1898.
16 Cited in Todd, p. 12.
17 Noted in ibid, p. 60.
18 Letter to May, 25 September 1898, noted in ibid p. 63.
19 Campbell, pp. 250–51.
20 Rutherford to Mary, [no date] September 1899, in Eve, p. 68.
21 Reeves, p. 43.
22 R. B. Owens, 'Thorium radiation', *The Lon lon, Edinburgh and Dublin Philosophical Magazine and Journal*, Vol. 48, July–December 1899, pp. 360–87, pp. 361–62.
23 ATL MS-Papers-1342-258, Marsden, Ernest (Sir), 1889–1970: Papers. Facts and particulars concerning the early life of Lord Rutherford; H.R. Robinson, 'Rutherford in Manchester 1907–1919', typescript.
24 Campbell, p. 254.
25 R.B. Owens, 'Thorium radiation', p. 361.
26 E. Rutherford and F. Soddy, 'The Cause and Nature of Radioactivity, Pt 1', *The London, Edinburgh and Dublin Philosophical Magazine and Journal of Science*, 6th Series, Volume 4, July–December 1902, pp. 372–73.
27 Ibid, p. 373.
28 E. Rutherford, 'A Radio-active Substance emitted from Thorium Compounds', *The London, Edinburgh and Dublin Philosophical Magazine and Journal of Science*, 5th Series, Vol. 49, January–June 1900, pp. 1–14, p. 6].

29 Ibid, p. 14.
30 E. Rutherford, 'Radioactivity produced in Substances by the Action of Thorium Compounds', *The London, Edinburgh, and Dublin Philosophical Magazine and Journal of Science*, 5th Series, Vol. 49, Jan–June 1900, pp. 161–92, esp. p. 161.
31 ATL MS-Papers 1342-258, Marsden, Ernest (Sir), Facts and particulars concerning the early life of Lord Rutherford'; Henry T. Tizard, 'The Rutherford Memorial Lecture: Lord Rutherford, his life and influence on chemistry', 8 April 1939.
32 '*Comptes Rendus de l'Académie des Sciences*', i.e. '*Reports from the Scientific Academy*'.
33 Melinda Baldwin, 'Ernest Rutherford's Ambitions', *Physics Today*, Vol. 75, No. 5, May 2021, p. 30.
34 Rutherford to Mary, 18 November 1899, quoted in Eve, p. 69.
35 Campbell, p. 257.
36 Rutherford to Mary, undated, quoted in Campbell, p. 238.
37 Rutherford to Mary, 31 December 1899, quoted in Eve, p. 70.
38 Eve, p. 70.
39 Passenger list, *Auckland Star*, 8 May 1900.
40 *Opunake Times*, 11 May 1900.
41 *Star* (Christchurch), 19 May 1900.
42 *Press* (Christchurch), 23 May 1900. He was interviewed the day before.
43 E. Rutherford and F. Soddy, 'On Radioactive Change', *The London, Edinburgh and Dublin Philosophical Magazine and Journal of Science*, 6th Series, No. 5, January–June 1903, p. 591.
44 E. Rutherford and H.T. Barnes, 'Heating Effect of the Radium Emanation', *The London, Edinburgh and Dublin Philosophical Magazine and Journal of Science*, 6th Series, Jan–June 1904, pp. 202–19, esp. p. 219. The paper was dated 22 December 1903 and prefaced in a letter to *Nature*, Vol. 38, p. 622, 29 October 1903.
45 Born, pp. 210–12.
46 *Press* (Christchurch), 23 May 1900.
47 Quoted in Campbell, p. 265–66.
48 Ibid, p. 261.
49 Ibid, 9 July 1900.
50 *Taranaki Herald*, 9 June 1900.
51 Wtu MS-Papers-1342-251, Marsden, Ernest (Sir), 1889–1970,: Papers, Original correspondence relating to Rutherford, Mary Newton to Martha Rutherford, 22 May 1900.
52 Campbell, p. 262.
53 Rutherford to his mother, 2 April 1901, quoted in Eve, p. 76.
54 Campbell, p. 263.
55 Rutherford to Thomson, 26 March 1901, cited in Eve p. 77.
56 Rutherford to Mary, [no date], September 1899, quoted in Eve, p. 68.
57 'Harriet Brooks', www.thecanadianencyclopedia.ca/en/article/harriet-brooks, accessed 26 November 2024.
58 Campbell, p. 265.
59 Frederick Soddy, 'Reminiscences of McGill, 1900–1902', *Old McGill*, Vol. 36, McGill University, Montreal, 1933, p. 19.
60 Noted in Campbell, p. 265.
61 Soddy, 'Reminiscences of McGill', p. 19.
62 archivalcollections.library.mcgill.ca/index.php/harrington-bernard-j-bernard-james-1848-1908, accessed 18 November 2024.
63 Ibid, p. 20.
64 Baldwin, 'Ernest Rutherford's Ambitions'.
65 E. Rutherford, 'Emanations from radio-active substances', *Nature*, 13 June 1901, pp. 157–58.
66 Pierre Radvanyi, 'The discussion between P. Curie and E. Rutherford (1900–1904)', *The European Physical Journal*, Vol. 38, 2013, pp. 433–41.
67 Eve, p. 75.
68 CUL RP Add 7653, PA 6A-6B, 'The Origins of Radium', hand-written draft [1907].
69 CUL RP Add 7653 PA 4A-5, 'Abstract for the American Physical Society, "Excited radioactivity and Ionization in atmosphere"', typescript. The original typescript read '... in atmospheric air', struck out. The unedited title was repeated on the first page without amendment.
70 E. Rutherford, 'The recent radium controversy', Letters, *Nature*, 25 October 1906.
71 Eve, p. 75.
72 CUL RP Add 7654, Correspondence R, Rutherford to Thomson, 9 January 1900.
73 E. Rutherford, 'A Radio-active substance emitted from Thorium Compounds', *The London, Edinburgh and Dublin Philosophical Magazine and Journal of Science*, January 1900, pp. 1–14.
74 Noted by Baldwin, p. 30.
75 CUL RP Add 7654, Correspondence R, Rutherford to Thomson, 9 January 1900.
76 See arxiv.org/. The concept of 'pre-print' implies the paper is also pre-peer review, i.e. it is unverified.
77 Rutherford dated the manuscript 30 May and the magazine issued it on 13 June. E. Rutherford, 'Emanations from radio-active substances', *Nature*, 13 June 1901, pp. 157–58.
78 Baldwin, pp. 30–31.
79 CUL RP Add 7653, Correspondence D1-81, Dana to Rutherford, 1 November 1906.
80 E. Rutherford, 'Emanations from radio-active substances', *Nature*, 13 June 1901, pp. 157–58.
81 See E. Rutherford, 'Emanations from radio-active substances', *Nature*, 13 June 1901, pp. 157–58; E. Rutherford and Miss H.T. Brooks, 'The New Gas from Radium', *Proceedings and Transactions of the Royal Society of Canada*, Series 7, Vol. 7, 1901. Section III, pp. 21–25.
82 E. Rutherford, 'Emanations from Radio-active substances', *Nature*, 13 June 1901.
83 John Campbell, correspondence to author 17 March 2025.
84 E. Rutherford, 'Einfluss der Temperatur auf die "Emanationen" Radioaktiver Substanzen', *Physikalische Zeitschrift*, No. 29, 20 April 1901.
85 E. Rutherford, 'Emanations from Radio-active substances', *Nature*, 13 June 1901.
86 In hindsight this was because Rutherford's thorium emanation was radon-220, with a half-life of 54.5 seconds, whereas other radon isotopes had variously longer or shorter half-lives.
87 E. Rutherford, 'Emanations from Radio-active substances', *Nature*, 13 June 1901.
88 Ibid.

89 Noted by James L. Marshall and Virginia R. Marshall, 'Ernest Rutherford: the "true discoverer" of Radon,' *Bulletin for the History of Chemistry*, Vol. 28, No. 2, 2003, pp. 76.
90 Rutherford reported it in E. Rutherford, 'A Radio-active Substance emitted from Thorium Compounds,' *The London, Edinburgh, and Dublin Philosophical Magazine and Journal of Science*, 5th Series, Vol. 49, January–June 1900.
91 E. Rutherford and Miss H.T. Brooks, 'The New Gas from Radium,' *Proceedings and Transactions of the Royal Society of Canada*, Series 7, Vol. 7, 1901. Section III, pp. 21–25.
92 E. Rutherford, 'Emanations from Radio-active substances,' *Nature*, 13 June 1901.
93 Ibid.
94 Jeremy C.T. Fairbank, 'William Adams and the spine of Gideon Algernon Mantell,' *Ann R. Coll Surg. England*, 2004, Vol. 686, pp. 349–52. Remarkably, Mantell's journal is held in New Zealand.
95 CUL RP Add 7653, Correspondence K1-61, Onnes to Rutherford, 9 December 1912.
96 CUL RP Add 7653, Correspondence C54-93, Cooke to Rutherford, 5 March 1904.
97 CUL RP Add 7653, Correspondence B417-465, Brown to Rutherford, 28 May 1906.
98 CUL RP Add 7653, Correspondence M1-54, McClurg to Rutherford, 11 December 1901.
99 ATL Micro MS-Coll-20-2016, Personal collection of Sir William Bragg (continued), Rutherford to William Bragg, 22 October 1904.
100 CUL RP Add 7653, Correspondence C54-93, Cooke to Rutherford, 2 November 1903.
101 Ibid, Child to Rutherford, 12 September 1901. This file included letters from Marie Curie and was labelled radioactive by the archive.
102 CUL RP Add 7653, Rutherford Correspondence B417-465, Brooks to Rutherford, 8 December 1901.
103 Reeves, p. 53.
104 CUL RP Add 7654, Correspondence R, Rutherford to Thomson, 26 [December?] 1902.
105 For summary of liquid air vs specific gases see http://www.hyperphysics.phy-astr.gsu.edu/hbase/thermo/liqair.html, accessed 20 December 2024.
106 Frederick Soddy, 'Reminiscences of McGill, 1900–1902.'
107 Ibid.
108 James Jeans, 'The Mechanism of Radiation,' *The London, Edinburgh and Dublin Philosophical Magazine and Journal of Science*, 6th Series, Vol. 2, Jul–Dec 1901, pp. 421–55.
109 CUL RP Add 7653, Correspondence C54-93, Crookes to Rutherford, 18 December 1901.
110 Rutherford to his mother, 5 January 1902, cited in Eve, p. 80.
111 Soddy, 'Reminiscences of McGill.'
112 E. Rutherford and F. Soddy, 'The cause and nature of radioactivity, Part 1,' *The London, Edinburgh and Dublin Philosophical Magazine and Journal of Science*, 6th Series, Vol. 4, July–December 1902, pp. 370–96.
113 Ibid, p. 395.
114 Ibid, p. 397.
115 CUL RP Add 7653, Correspondence C54-93, Rutherford to Crookes, 29 April 1902.
116 Rutherford to his mother, 1 August 1902, cited in Eve p. 83.
117 E. Rutherford and F. Soddy, 'On Radioactive Change,' *The London, Edinburgh and Dublin Philosophical Magazine and Journal of Science*, 6th Series No. 5, January–June 1903, pp. 576–91, p. 591.
118 E. Rutherford 'Some Remarks on Radioactivity,' *The London, Edinburgh and Dublin Philosophical Magazine and Journal of Science*, 6th Series No. 5, January–June 1903, pp. 480–85.
119 CUL RP, Correspondence L76-146, Lodge to Rutherford, 11 December 1903.
120 ATL Micro MS-Coll-20-2016, Personal collection of Sir William Bragg (continued), Rutherford to William Bragg, 3 July 1903.
121 E. Rutherford and H.T. Barnes, 'Heating Effect of the Radium Emanation,' Letters, *Nature*, 29 October 1903, p. 622.
122 Ernest Rutherford, *Radioactivity*, Cambridge University Press, Cambridge, 1904.
123 See, e.g. CUL RP Add 7653, Correspondence, A1-72, S.J. Allan to Rutherford, 19 May 1903. This process still took time: Rutherford returned the paper in August, see Allan to Rutherford, 16 August 1903.
124 Todd, pp. 14–15, quoting Soddy, original source unstated.
125 Lawrence Badash (ed.), *Rutherford and Boltwood: letters on radioactivity*, Yale University Press, New Haven, 1969, Boltwood to Rutherford, 8 August 1904, pp. 36–37. Block capitals in original.
126 Todd, p. 16.
127 For example, Rutherford's laboratory notebooks in CUL MS-Add 7653, Laboratory Notebooks NB 7-16, which span his late McGill and early Manchester days, are still radioactive to the point where for reader safety they are marked with an international radiation symbol both on the exterior holding box and on the envelopes within, with warnings requiring readers to wear gloves, as are other files in the NB-series notebook sequence.
128 Noted in Eve, p. 138.
129 Noted in ibid, p. 111.
130 CUL RP Add 7653, Correspondence B1-53, Barnard to Rutherford, 22 May 1905. Barnard is perhaps best known for his discovery of Barnard's Star in 1916.
131 CUL RP Add 7653, Correspondence L1-75, Rutherford to Laird, 5 November 1904.
132 Eve, p. 113.
133 Marjorie Malley, 'The discovery of atomic transmutation: scientific styles and philosophies in France and Britain,' *Isis*, Vol. 70, No. 2, June 1979. [Note: www.journals.uchicago.edu/doi/10.1086/352196]
134 Roberto de Andrade Martins, *Historical Studies on Radioactivity*, Quamcumque Editum, 2021, pp.167–92, paper presented 2005. [Publication location not given.]
135 ATL Micro MS-Coll-20-2016, Personal collection of Sir William Bragg (continued), Rutherford to Bragg, 22 September 1907.

7 A mass of energy

1 Noted in Eve, p. 95.
2 E. Rutherford, 'The recent radium controversy,' Letters, *Nature*, 25 October 1906, pp. 634–35.
3 Noted in Eve, p. 98.

4 Quoted in ibid, p. 129.
5 Ernest Rutherford, *Radioactive Transformations*, Yale University Press, New Haven, 1906, p. 186.
6 James Dewar, 'The rate of production of helium from radium', *Proceedings of the Royal Society of London*, Series A, Vol. 81, Issue 507, September 1908, p. 285.
7 CUL RP Add 7653, Misc. files and notes, PA312/3, 'To the editor Nature — Hahn', typescript.
8 ATL MS-Papers-1342-260 Marsden, Ernest (Sir), 1889–1970, papers, 'Addresses and articles on Lord Rutherford', Otto Hahn, 'Extracts from "The New Atoms"', typescript.
9 Rutherford to Boltwood, 9 April 1905, in Badash (ed.), p. 56.
10 ATL Micro MS-Coll-2-2016, Personal collection of Sir William Bragg (continued), Rutherford to William Bragg, 14 November 1904. Not known whether US or Canadian dollars.
11 Ibid, Rutherford to Bragg, 14 November 1904. The microfilm copy is significantly degraded.
12 Badash (ed.), Rutherford to Boltwood, 28 April 1905, p. 69.
13 Ibid, Rutherford to Boltwood, 15 April 1905; Boltwood to Rutherford, 18 April 1905, etc, pp. 62–67.
14 CUL RP Add 7653, Correspondence, K1-61, Kelvin to Rutherford, 16 October 1906.
15 *Auckland Star*, 7 June 1905.
16 Ibid. This report was widely syndicated, e.g. *Hawera and Normanby Star*, 9 June 1905, *Otago Witness*, 14 June 1905,
17 Quoted in Eve, p. 133.
18 Ernest Rutherford, *Radioactive Transformation*, Yale University Press, New Haven, 1906.
19 See, e.g. Oliver Lodge letter to *The Times*, 4 September 1906.
20 E. Rutherford, 'The Recent Radium Controversy', *Nature*, 25 October 1906.
21 E. Rutherford, 'The Recent Radium Controversy', *Nature*, 25 October 1900, pp. 634–35.
22 E. Rutherford and F. Soddy, 'On Radioactive Change', p. 591.
23 John Draper, 'On the production of light by heat', *The London, Edinburgh and Dublin Philosophical Magazine and Journal of Science*, Series 3, Vol. 30, July–December 1847, pp. 345–60.
24 Calculating the actual volume of a complex three-dimensional object with multiple non-symmetrical compound curves and no symmetrical axis of revolution such as a cow is extraordinarily challenging, but once assumed to be a sphere, the volume of a cow is calculable via the usual formula for calculating spherical volume, v = 4/3 πr3, where v is volume and r radius.
25 This model is known as ACDM; for analysis of the structure see Antonia Siefert, Zachary G. Lane, Marco Galoppo, Ryan Ridden-Harper, David L. Wiltshire, 'Supernovae evidence for foundational change to cosmological models', *Monthly Notices of the Royal Astronomical Society: Letters*, Vol. 537, Issue 1, February 2025, pp. L55–L60, doi.org/10.1093/mnrasl/slae112. academic.oup.com/mnrasl/article/537/1/L55/7926647.
26 A useful summary is given in Himanshu Mavani and Navinder Singh, 'A concise history of the black-body radiation problem', arxiv.org/pdf/2208.06470, April 2022, esp. pp. 8–9.
27 Max Planck 'Ueber das Gesetz der Energieverteilung im Normalspectrum', *Annalen der Physik*, Vol. 309, No. 3, 1901, pp. 553–63.
28 This has been extensively covered in the literature but see, e.g. Martin J. Klein, 'Max Planck and the beginnings of quantum theory', *Archive for the History of Exact Sciences*, Vol. 1, No. 5, 1962, pp. 459–79.
29 Specifically, Rutherford provided a value for the charge on an electron that matched the value required to validate Planck's equations. See Ernest Rutherford, 'The development of the theory of atomic structure', in Joseph Needham and Walter Pagel (eds), *Background to Modern Science*, Cambridge University Press, Cambridge, 1938, pp. 64–65. The book contained ten lectures dating from 1936. Rutherford's contributions were published posthumously, edited by J.A. Ratcliffe.
30 See e.g. CUL Add 7653, RP, Notes and Manuscripts, PA194 (part), 'Theory of Structure of Atom', working notes, pp. 1–4 and 17–18, etc.
31 J.J. Thomson, 'On the Structure of the atom', *The London, Edinburgh and Dublin Philosophical Magazine and Journal of Science*, Vol. 7, No. 49, Jan–June 1904, pp. 237–65, esp. pp. 238–53.
32 *Rick and Morty* (Adult Swim, 2013), apparently originating as parody of *Back to the Future* (Universal, 1985), which also featured a scientist with Einstein-like hair.
33 For an English version see A. Einstein, 'Concerning the Investigation of the State of Aether in Magnetic Fields', c.1894, 1library.net/document/y9jgg4vq-concerning-the-investigation-of-the-state-of-aether-in-magnetic-fields-by-albert-einstein.html, accessed 13 November 2024.
34 Norbert Straussman, 'On Einstein's Doctoral Thesis', www.researchgate.net/publication/2172866_On_Einstein's_Doctoral_Thesis, accessed 13 November 2024. The reason why Avogadro's number is so crucial to chemistry is because it defines the number of particles in a given quantity of a substance, following Avogadro's Law in which gases of the same temperature and pressure contain the same number of particles. Avogadro's number became a means for relative comparison. Putting an actual value to it — that is, defining the actual number of particles — enabled a *quantitative* value to be given.
35 For useful summary see Jozeph T. Devreese, 'The light came in 1905', arxiv.org/pdf/physics/0602083, accessed 1 January 2025.
36 A. Einstein, 'Zur Elektrodynamik bewegter Körper', *Annalen der Physik*, Vol. 322, No. 10, 1905, pp. 891–21. For English translation see A. Einstein, 'On the electrodynamics of moving bodies', reproduced in W. Perrett and G.B. Jeffery (trans.), *The Principle of Relativity*, Dover Publications, reprint, Methuen & Co., 1923, pp. 35–65.
37 A. Einstein, '1st die Tragheit eines Korpers von seinem Energiegehalt abhangig?', *Annalen der Physik*, 17, 1905. For English translation see A. Einstein, 'Does the inertia of a body depend on its energy content?', reproduced in W. Perrett and G.B. Jeffery (trans.), *The Principle of Relativity*, Dover Publications, reprint, Methuen & Co., 1923, pp. 69–71.
38 For example, Born, p. 283.
39 English translation at: einsteinpapers.press.princeton.edu/vol2-trans/186, accessed 1 January 2025.
40 Software developed by MIT plays with this concept by progressively reducing the speed of light, making it

possible to see in human terms what would happen if Einstein's relativistic effects applied at everyday velocities: http://gamelab.mit.edu/games/a-slower-speed-of-light/, accessed 4 January 2025.

41 In SI units, e=mc2 produces 629,128,625,115,772 joules from 7 grams of any matter, 1 kilowatt-hour is equal to 3.6 million joules.

42 Total generation was 43,488,000 kw/h, figures via MBIE, www.mbie.govt.nz/building-and-energy/energy-and-natural-resources/energy-statistics-and-modelling/energy-statistics/electricity-statistics, accessed 3 January 2025.

43 energyeducation.ca/encyclopedia/Nuclear_fusion_in_the_Sun, accessed 3 January 2025.

8 Journey to the centre of the atom

1 Alexander Turnbull Library, ATL MS-Papers 1342-258, Marsden, Ernest (Sir), Facts and particulars concerning the early life of Lord Rutherford', C.D. Ellis, 'The Cavendish Chair', typescript.

2 CUL RP Add 7653, Correspondence H1-88, Rutherford to Hahn, 21 October 1906, draft typed version with pen-and-ink emendations. See also Marckwald to Rutherford, 19 July 1906 in ibid M1-54, spelt 'Rutherfordit'. The mineral, now known as 'Rutherfordine', forms brownish-yellow crystals.

3 Clifford Frondel and Robert Meyrowitz, 'Studies on uranium minerals: Rutherfordine, Diderchite and Clareite', Trace Elements Investigations Report 474, United States Department of the Interior Geological Survey, October 1954, p. 6.

4 Rutherford, *Radioactivity*, p. 1.

5 CUL RP Add 7653, Correspondence H1-88, Rutherford to Hahn, 20 August 1906.

6 Rutherford to his mother, 10 July 1906, quoted in Eve, p. 147.

7 ATL MS-Papers-1342-252, Marsden, Ernest (Sir) (1889–1970), Paper: Correspondence — Lord Rutherford, Rutherford to Schuster, 26 September 1906.

8 CUL RP Add 7653, Correspondence S1-88, Schuster to Rutherford, 7 September 1906.

9 ATL MS-Papers-1342-252, Marsden, Ernest (Sir) (1889–1970), Paper: Correspondence — Lord Rutherford, Rutherford to Schuster, 26 September 1906.

10 Ibid, Rutherford to Schuster, 26 September 1906.

11 CUL RP Add 7653, Correspondence S1-88, Schuster to Rutherford, 7 October 1906. This letter and Rutherford's of 26 September were published in J.B. Birks (ed.), *Rutherford at Manchester*, Heywood & Company, London, 1962, pp. 47–52; however, the archival files show the correspondence was more extensive.

12 CUL RP Add 7653, Correspondence S1-88, telegram, 8 November 1907.

13 www.bankofengland.co.uk/monetary-policy/inflation/inflation-calculator, accessed 5 February 2025.

14 CUL RP Add 7653, Correspondence P1-50, Rutherford to Peterson, 1 January 1907 (draft typescript with pencil emendations).

15 ATL Micro-MS-Coll-20-2015, Records of Secretary: Personal collections of Sir Humphry Davy, Michael Faraday, Sir William Grove, John Tyndall and Sir William Bragg; Soddy to Bragg, 10 January 1907.

16 CUL RP Add 7653, Correspondence N-O, Nichol to Rutherford, 6 February 1907.

17 Ibid, Nichol to Rutherford, 26 May 1907.

18 CUL RP Add 7653, Correspondence P1-50, Rutherford to Peterson, 1 January 1907 (draft typescript with pencil emendations).

19 Badash (ed.), Rutherford to Boltwood, 10 November 1906, p. 145.

20 — *The Physical Laboratories of Manchester: a record of 25 Years' work*, Manchester University Press, Manchester, 1906, p. 1.

21 Ibid, p. 3.

22 — *The Physical Laboratories of Manchester: a record of 25 Years' work*, p. 1.

23 Campbell, pp. 302–303.

24 F.R. Terroux, 'The Rutherford Collection of Apparatus at McGill University', *Transactions of the Royal Society of Canada*, S. III, 1938, p. 9, reprint in CUL RP Add 7653, Misc. files and notes, PA 315.

25 ATL Micro MS-Coll-20-2016, Personal collection of Sir William Bragg (continued), Rutherford to Bragg, 12 March 1907.

26 CUL RP Add 7653, Correspondence G1-80, Geiger to Rutherford, 13 May 1907. The date of this letter suggests it probably pursued him across the Atlantic to Manchester.

27 Noted in Eve, pp. 157–58.

28 ATL Micro MS-Coll-20-2016, Personal collection of Sir William Bragg (continued), Rutherford to William Bragg, 5 July 1907.

29 Noted in Todd, p. 17.

30 Badash (ed.), Rutherford to Boltwood, 28 July 1907, p. 159.

31 Actinium is a highly radioactive element with the atomic number of 89, discovered in 1899 in very small amounts from uranium ores.

32 Ibid, Rutherford to Boltwood, 28 July 1907, p. 158.

33 CUL RP Add 7653, Laboratory Notebooks, NB7-16, NB 12, Tuesday 8 October 1907.

34 Todd, p. 20.

35 CUL RP Add 7653, Rutherford Correspondence R1-74, Ramsay to Rutherford, 11 November 1907.

36 Ibid, Rutherford to Ramsay, 13 November 1907.

37 Ibid, Ramsay to Rutherford 14 December 1907.

38 Badash (ed.), Rutherford to Boltwood, 15 February 1908, p. 180.

39 CUL RP Add 7653, Notes and Manuscripts, PA35-37, Lecture, Royal Institution, 24 May 1913.

40 CUL RP Add 7653, Notes and Manuscripts, PA190-193.

41 Rutherford to Schuster, 26 September 1906, in Eve, pp. 144–45.

42 H.R. Robinson, 'Rutherford: Life and work to the year 1919, with personal reminiscences of the Manchester period', in Birks (ed.), p. 75.

43 See, e.g. CUL RP Add 7653, Correspondence , K1-61, Kaye to Rutherford, 1 July 1913.

44 CUL RP Add 7653, Correspondence F1-66, see, e.g. Fajans to Rutherford, 7 June 1912 (English), Fajans to Rutherford, 9 January 1913 (German), etc.

45 Ibid, Correspondence B417-465, Bumstead to Rutherford, 23 June 1908.

46 Eve, p. 166, citing Rutherford to Boltwood, 30 October 1907.

47 ATL Micro MS-Coll-20-2016, Personal collection of Sir William Bragg (continued), Rutherford to Bragg, 10 November 1907.

48 Ibid, Bragg to Rutherford, 22 January 1908.
49 See, e.g. CUL RP Add 7653, correspondence M1-54, McLelland to Rutherford, 8 February 1908.
50 Noted in Eve, p. 158.
51 CUL RP Add 7653, Rutherford Correspondence B54-96, Bickerton to Rutherford 24 December 1908 (etc); see also ATL Micro-MS-Coll-20-1939, Larmor, Lloyd, Lubbock and RP, Bickerton to Rutherford, 21 October 1910; 17 November 1910; 19 November 1910, 6 January 1911, 11 January 1911 (etc).
52 CUL RP Add 7653, Correspondence I-J, Innes to Rutherford, 29 January 1909.
53 Ibid, C54-115, Rutherford for Curie, 9 May 1910.
54 CUL RP Add 7653, Misc. files and notes, PA306, Newspaper cuttings and letters of E.R. to Mother (published by Jim R.), Rutherford to his mother, 14 October 1910.
55 E. Rutherford, 'The international radium standard', *Nature*, 4 April 1912.
56 CUL, Add-7653, RP, Correspondence C54-115, note in French.
57 Ibid, M102-160, Rutherford to Meyer, 8 September 1910.
58 Museum of Radiation and Radioactivity, 'How the Curie Came to Be', orau.org/health-physics-museum/articles/how-the-curie-came-to-be.html, accessed 11 February 2025.
59 CUL RP Add 7653, Misc. files and notes, PA306, Newspaper cuttings and letters of E.R. to Mother (published by Jim R.), Rutherford to his mother, 14 October 1910.
60 CUL RP Add 7653, Misc. files and notes, PA361, Lists of Projected Researches, 'researches for 1908–1909 (projected).
61 E. Rutherford, H. Geiger, 'An electrical method of counting the number of α-particles from radio-active substances', *Proceedings of the Royal Society of London*, Series A, Vol. 81, Issue 546, Aug 1908, pp. 141–62.
62 Rutherford to Hahn, 27 January 1908 in Eve, p. 175.
63 E. Rutherford, H. Geiger, 'The charge and nature of the α-particle', *Proceedings of the Royal Society of London*, Series A, Volume 81, Issue 546, Aug 1908, pp. 162–73, esp. p. 172.
64 Rutherford to Bumstead, 11 July 1908, in Eve, p. 180.
65 At New Zealand's average domestic line value of 240 volts, a two-kilowatt heater draws 8.33 amperes.
66 E. Rutherford, H. Geiger, 'The charge and nature of the α-particle', p. 172. Today we know it is a helium atom minus its two electrons, i.e. a helium nucleus, but Rutherford had yet to discover that in 1908.
67 Robinson, in Birks (ed.), p. 73.
68 www.nobelprize.org/nomination/archive/list.php, accessed 22 February 2025.
69 Ibid.
70 www.nobelprize.org/prizes/chemistry/1908/rutherford/facts/, accessed 22 February 2025.
71 Rutherford to Hahn, 29 November 1908, in Eve, p. 183.
72 See, e.g. *Evening Star*, 25 November 1908.
73 Compare, e.g. *Fielding Star*, 25 November 1908 with *Evening Star*, 25 November 1908.
74 *Taranaki Daily News*, 28 November 1908.
75 *Colonist*, 26 November 1908.
76 *Press*, 26 November 1908.
77 *Evening Post*, 28 November 1908.
78 *Manawatu Times*, 16 December 1908.
79 Campbell, p. 313. This song, a popular ditty from the comic opera 'La Poupée', was written by Edmond Audran.
80 Professor K.B. Hasselberg, 'Award Ceremony Speech', www.nobelprize.org/prizes/chemistry/1908/ceremony-speech/, accessed 6 January 2025.
81 Eve, p. 183.
82 CUL RP Add 7653, Correspondence R1-74, Rayleigh to Rutherford, 22 December 1908.
83 CUL RP Add 7653, Misc. files and notes, PA306, Newspaper cuttings and letters of E.R. to Mother (published by Jim R.), Rutherford to his mother, 24 December 1908
84 www.bankofengland.co.uk/monetary-policy/inflation/inflation-calculator; matching 1908 against December 2024 CPI values.
85 CUL RP Add 7653, Misc. files and notes, PA306, Newspaper cuttings and letters of E.R. to Mother (published by Jim R.), Rutherford to his mother, 6 April 1910.
86 H. Geiger and E. Marsden, 'On a diffuse reflection of the α-particles', *Proceedings of the Royal Society of London*, Series A, Vol. 82, No. 557, July 1909.
87 CUL RP Add 8653, Rutherford Correspondence C1-53, C8, Callendar to Rutherford, 7 June 1910.
88 Noted in E. Rutherford, 'The scattering of α and β particles by matter and the structure of the atom', in Ernest Rutherford, *The Collected Papers of Lord Rutherford of Nelson, Volume 2: Manchester*, Routledge, New York, 2016 (paperback), pp. 238–39.
89 For discussion of this and other early atomic models see Charles Bailly, 'Atomic modelling in the early 20th century: 1904–1913', University of Colorado, Boulder, Department of Physics, October 12, 2008, http://philsci-archive.pitt.edu/4232/1/Atomic_Modeling_101208.pdf, esp. pp. 6–7.
90 CUL RP Add 7653, N-O, Nagaoka to Rutherford, 21 January 1911.
91 Eve, p. 198.
92 Conic geometry describes parabolas, and the mathematics to do so are simpler than Newton's gravitation equations. Orbital simulation software such as Kerbal Space Program use conics as an alternative to n-body physics because of the reduced load on the computer.
93 E. Rutherford, 'The scattering of α and β particles by matter and the structure of the atom', in Ernest Rutherford, *The Collected Papers of Lord Rutherford of Nelson, Volume 2: Manchester*, Routledge, New York, 2016 (paperback), p. 254.
94 E. Rutherford, 'The scattering of α and β particles by matter and the structure of the atom', in Ernest Rutherford, *The Collected Papers of Lord Rutherford of Nelson, Volume 2: Manchester*, Routledge, New York, 2016 (paperback), pp. 240–54.
95 E. Rutherford, 'The structure of the atom', *The London, Edinburgh and Dublin Philosophical Magazine and Journal of Science*, 6th Series, Vol. 27, Jan–June 1914, pp. 488–98, esp. p. 488.
96 CUL RP Add 7653, Correspondence G1-80, Geiger to Rutherford, 13 October 1912.
97 CUL RP Add 7653, Notes and Manuscripts PA194

(part), 'Theory of structure of atom'; continued in a second folder with discontiguous page numbering. These are undated but likely were preliminary workings ahead of his formal papers.

98 CUL RP Add 7653, Correspondence P1-80, Charles Phillips to Rutherford, 8 July 1912.

99 CUL RP Add 7653, Correspondence B170-229, Boltwood to Rutherford, 2 November 1910.

100 Noted in Fritz Berends and Franklin Lambert, 'Einstein's Witches', *Europhysics News*, July 2018, paperity.org/p/100465520/einsteins-witches, accessed 13 January 2025.

101 www.solvay.com/en/article/inventing-soda-ash-again, accessed 13 January 2025.

102 Solvay to Poincaré, invitation letter 15 June 1911, published at http://numericana.com/fame/invitation.htm, accessed 13 June 2025.

103 Niels Bohr, 'The Solvay Meetings and the Development of Quantum Mechanics', Niels Bohr at the occasion of the 12th Solvay Conference in physics, 9–14 October 1961, http://www.solvayinstitutes.be/pdf/Niels_Bohr.pdf accessed, 13 January 2025.

104 P. Langevin and M. de Broglie, *La Theorie du Rayonnement et les Quanta, rapports et discussions*, Guthier-Villars, Paris 1912, pp. 388–90.

105 CUL RP Add 7653, Correspondence B417-465, de Broglie to Rutherford, 23 December 1911.

106 Calculateur d'inflation, france-inflation.com/calculateur_inflation.php, accessed 15 January 2025.

107 — *La structure de la Matière, Rapports et discussions du Conseil de Physique*, Gauthier-Villars, Paris, 1921, pp. vii-viii. This volume compiled the reports on the 1913 conference but did not appear until after the First World War.

108 explore.psl.eu/en/magazine/focus/introduction-solvay-conferences-physics, accessed 14 January 2025.

109 P. Langevin and M. de Broglie, *La Theorie du Rayonnement et les Quanta, rapports et discussions*, Guthier-Villars, Paris, 1912, pp. 388–90.

110 Niels Bohr, 'The Solvay Meetings and the Development of Quantum Mechanics', Niels Bohr at the occasion of the 12th Solvay Conference in physics, 9–14 October 1961, http://www.solvayinstitutes.be/pdf/Niels_Bohr.pdf, accessed 13 January 2025.

111 CUL RP Add 7653, Rutherford Correspondence B97-169, Bohr to Rutherford, 5 November 1912; Bohr to Rutherford, 31 January 1913; Bohr to Rutherford, 6 March 1913.

112 Ibid, Bohr to Rutherford, 6 March 1913.

113 Ibid, Bohr to Rutherford, 8 April 1913.

114 Eve, p. 218.

115 N. Bohr, 'On the Constitution of Atoms and Molecules', *The London, Edinburgh and Dublin Philosophical Magazine and Journal of Science*, 6th Series, Vol. 26, July–December 1913, p. 1.

116 CUL RP Add 7653, Misc. files and notes, PA 363-374, 'Professor Niels Bohr', typescript.

117 CUL RP Add 7653, Correspondence B97-169, Rutherford to Bohr, 20 March 1913.

118 See, e.g. ibid, Bohr to Rutherford, 31 December 1913.

119 CUL RP Add 7653, Misc. files and notes, PA 363-374, 'Professor Niels Bohr', typescript.

120 — *La Structure de la Matière, rapports et discussions du Conseil de physique tenu à Bruxelles du 27 au 31 Octobre 1913, sous les auspices de l'Institut international de physique Solvay*, Gauthier-Villars et Co., Paris, 1921, p. 53. Production of this volume was interrupted by war.

121 Ibid, pp. 53–55.

122 Ibid, pp. 45–74

123 E. Rutherford, 'International conference on the structure of matter', *Nature*, 20 November 1913, p. 347.

124 CUL RP Add 7653, Correspondence P1-50, Pope to Rutherford, 18 November 1913.

125 E. Rutherford, 'The structure of the atom', p. 490.

126 Ibid, n p. 498.

127 Ibid, esp. p. 488.

9 Disintegrating atoms

1 CUL RP Add 7653, Correspondence H89-125, Hale to Rutherford, 1 June 1914. The reference to trench warfare is interesting: the First World War had yet to break out, but the principles were well known.

2 CUL RP Add 7653, Notes and manuscripts, Misc. Xerox (box), Rutherford to Hevesy, 8 January 1914.

3 CUL RP Add 7653, Correspondence R75-137, telegram, 2 January 1914.

4 See, e.g. *Manawatu Standard*, 2 January 1914; *Press*, 2 January 1914, *New Zealand Times*, 2 January 1914, etc.

5 *Otago Witness*, 7 January 1914.

6 *Southland Times*, 3 January 1914.

7 *Evening Star*, 2 January 1914.

8 *Colonist*, 3 January 1914.

9 *Fielding Star*, 2 January 1914.

10 H.G.J. Moseley, 'The High-Frequency Spectra of the Elements', *The London, Edinburgh and Dublin Philosophical Magazine and Journal of Science*, 6th Series, Vol. 26, July–December 1913, pp. 1024–34.

11 CUL RP Add 7653 Correspondence M238-264, Moseley to Rutherford, n.d. (1913), Moseley to Rutherford, 5 January 1914; Moseley to Rutherford, 4 March 1914, etc.

12 E. Rutherford, 'The constitution of matter and the evolution of the elements', *Popular Science Monthly*, Vol. 87, August 1915, pp. 105–42.

13 Ibid, p. 136.

14 Ibid, p. 142.

15 CUL RP Add 7653 Correspondence H89-125, Hale to Rutherford, 1 June 1914.

16 Ibid, Hall-Jones to Rutherford, 24 August 1911; Hall-Jones to Rutherford, 28 August 1911; Hall-Jones to Rutherford, 29 August 1911; W. A. Heedman to Rutherford, 29 August 1911. William Hall-Jones was the New Zealand High Commissioner in London.

17 CUL RP Add 7653, Correspondence M161-233, Rutherford to Meyer, 29 June 1914.

18 Campbell, p. 350.

19 http://ssmaritime.com/SS-Euripides.htm, accessed 16 January 2025.

20 Noted in Campbell, p. 353.

21 See Matthew Wright, *The New Zealand Experience in Gallipoli and the Western Front*, Oratia Books, Auckland, 2017, pp. 52–56. Von Spee steamed for the Carolines, then Sāmoa, then Chile, the latter a neutral ground where he could obtain coal.

22 H.G. Wells, 'The World Set Free', www.gutenberg.org/cache/epub/1059/pg1059-images.html#chap01, accessed 7 January 2025.

23 This interview was syndicated, e.g. *Patea Mail*, 7 September 1914.

24 *Sun* (Christchurch), 14 October 1914.
25 ATL Micro-MS-Coll-20-1939, Larmor, Lloyd, Lubbock and Rutherford Papers, Rutherford to Boltwood, 29 October 1914.
26 Ibid. Rutherford clearly knew about the issues ships faced with marine fouling when out of dock for extended periods.
27 Geoffrey Bennett, *Coronel and the Falklands*, Pan, London, 1962, pp. 13–42. Bennett argues that this battle occurred due to Admiralty meddling from London, enabled by their new wireless communications network. This was the system envisioned when Rutherford initially worked in wireless in the 1890s.
28 ATL Micro-MS-Coll-20-1939, Larmor, Lloyd, Lubbock and Rutherford Papers, Rutherford to Boltwood, 14 September 1915.
29 CUL RP Add 7653, Correspondence M161-233, Meyer to Rutherford, 12 March 1915; Meyer to Rutherford, 3 September 1915.
30 CUL RP Add 7653, Correspondence G1-80, Geiger to Rutherford, 26 March 1915.
31 See, e.g. CUL RP Add 7653, Correspondence C1-53, Chadwick to Rutherford, 14 September 1915; Geiger to Rutherford, 24 May 1918.
32 Ibid, Chadwick to Rutherford, 31 March 1917
33 CUL RP Add 7653, Correspondence A1-72, Andrade to Rutherford, 15 June 1915.
34 CUL RP Add 7653, Correspondence D1-81, Darwin to Rutherford, 24 February 1917.
35 Letter series in CUL RP Add 7653, Correspondence A1-A72.
36 CUL RP Add 7653, Correspondence M55-101a, Marsden to Rutherford, 7 December 1914, 21 March 1915, 1 April 1915.
37 Ibid, Marsden to Rutherford, 18 September 1915.
38 Noted in Eve, pp. 253–54.
39 For summary details see www.mhs.ox.ac.uk/moseley/2015/08/10/100-years-later-the-life-and-death-of-henry-harry-moseley/index.html, accessed 17 January 2025.
40 CUL RP Add 7653 Correspondence M238-264, Moseley to Rutherford, 4 April 1915.
41 CUL RP Add 7653, Misc. files and notes, PA 375, Rutherford to Urbain, 26 September 1915.
42 CUL RP Add 7653, Misc. files and notes, PA375, reference for Moseley, 9 June 1914.
43 Cambridge University Press, RP, Add 7653, Correspondence A1-A72, Andrade to Rutherford, 20 September 1915.
44 E. Rutherford, 'Henry Gwyn Jeffreys Moseley', *Nature*, 9 September 1915, p. 33.
45 ATL Micro-MS-Coll-20-1939, Larmor, Lloyd, Lubbock and Rutherford Papers, Rutherford to Boltwood, 14 September 1915.
46 Roy M. MacLeod and E. Kay Andrews, 'Scientific advice in the War at Sea, 1915–1917: the Board of Invention and Research', *Journal of Contemporary History*, Vol. 6, No. 2, 1971, pp. 3-40, esp. p. 5.
47 Discussed in Matthew Wright, *Those Who Have the Courage*, Oratia Books, Auckland, 2024, pp. 154–56.
48 For discussion see Michael Pattison, 'Scientists, Inventors and the Military in Britain 1915–19: the Munitions Inventions Department', *Social Studies of Science*, Vol. 13, 1981, pp. 521–68.
49 Fisher to Balfour, 7 July 1915, in Arthur Marder (ed.), *Fear God and Dread Nought: the Correspondence of Admiral of the Fleet Lord Fisher of Kilverstone*, Vol. III, Jonathan Cape, London, 1959, p. 277.
50 Lord Fisher, *Records*, Hodder & Stoughton, London, 1919, pp. 65–66.
51 MacLeod and Andrews, p. 10.
52 Ibid, p. 13.
53 Katzir, p. 145.
54 ATL Micro-MS-Coll-20-1939, Larmor, Lloyd, Lubbock and Rutherford Papers, Rutherford to Boltwood, 21 February 1916.
55 Ibid, Rutherford to Boltwood, 21 February 1916.
56 CUL RP Add 7653, Rutherford Correspondence K1-61, King to Rutherford, 5 May 1916.
57 Eve, p. 268.
58 CUL RP Add 7653, Rutherford Correspondence T55-96, Threlfall to Rutherford, 5 October 1916, reporting on a visit by Threlfall to Jellicoe.
59 Shaul Katzir, 'Who knew piezoelectricity? Rutherford and Langevin on submarine detection and the invention of sonar', *Notes and Records of the Royal Society*, Vol. 66, 2012, pp. 141–57, esp. p. 143.
60 The idea that it was named after the originating committee is incorrect and apparently issued by the Admiralty in 1939. See www.historyofinformation.com/detail.php?entryid=1672, accessed 17 January 2025.
61 Katzir, pp. 150–51.
62 Eve, p. 253.
63 CUL RP Add 7653, Misc. files and notes, PA381-394, Llewellyn-Smith to Vice-Chancellor, The Victoria University, Manchester, 24 January 1916.
64 CUL RP Add 7653, Correspondence K1-61, King to Rutherford, 5 May 1916.
65 Ibid, Kaye to Rutherford, 18 October 1916.
66 John Gribbin, *Einstein's Master-Work: 1915 and the General Theory of Relativity*, Icon Books, London, 2015, pp. 150–54.
67 Discussed in Galina Weinstein, 'Einstein, Schwarzchild, the Perehelion Motion of Mercury and the Rotating Disk story', Tel Aviv University, 25 November 2014, pp. 5–7.
68 edition.cnn.com/2019/04/10/world/black-hole-photo-scn/index.html, accessed 21 January 2025.
69 CUL RP Add 7654, Rutherford Correspondence R, Rutherford to 'Professor' [Thomson], 4 May 1917.
70 CUL RP Add 7653, Misc. files and notes, PA306, Newspaper cuttings and letters of E.R. to Mother (published by Jim R.), Rutherford to his mother, 15 May 1917.
71 Rutherford to Mary Rutherford, 18 May 1917, quoted in Eve, p. 257.
72 CUL RP Add 7654, Correspondence R, Rutherford to 'Professor' [Thomson], 4 June 1917.
73 CUL RP Add 7653, Rutherford Correspondence L1-75, Langmuir to Rutherford, 18 May 1919.
74 Eve, p. 262.
75 Ruddock Mackay, *Fisher of Kilverstone*, Oxford University Press, Oxford, 1973, p. 506.
76 See MacLeod and Andrews, pp. 29–30.
77 CUL RP Add 7653, PA361, Lists of Projected Researches, 'Suggested notes of researches, March 15th 1915'.
78 E. Rutherford, 'On the Collision of α particles with Light Atoms, I, Hydrogen', *The London, Edinburgh and*

Dublin Philosophical Magazine and Journal of Science, 6th Series, Vol. 37, January–June 1919, pp. 537–61, p. 538.
79 Hydrogen is described as 'diatomic' because its atoms come in pairs, forming a molecule, that is, H2.
80 E. Rutherford, 'On the Collision of α particles with Light Atoms, I, Hydrogen', pp. 542–43.
81 CUL RP Add 7653, Notebooks, NB 17–25, NB24, 18 June 1918.
82 Sir Mark Oliphant, reading notes by Chadwick, 'Some personal recollections of Rutherford, The Man', *Notes and Records of the Royal Society of London*, Vol. 27, Issue 1, August 1972, pp. 7–23, p. 8.
83 CUL RP Add 7653, Misc. files and notes, PA306, Newspaper cuttings and letters of E.R. to Mother (published by Jim R.), Rutherford to his mother, 3 November 1918.
84 E. Rutherford, 'On the Collision of α particles with Light Atoms, I, Hydrogen', *The London, Edinburgh and Dublin Philosophical Magazine and Journal of Science*, 6th Series, Vol. 37, January–June 1919, pp. 537–61; E. Rutherford, 'On the Collision of α particles with Light Atoms, II, Velocity of the hydrogen atom', *The London, Edinburgh and Dublin Philosophical Magazine and Journal of Science*, 6th Series, Vol. 37, January–June 1919, pp. 562–70; E. Rutherford, 'On the Collision of α particles with Light Atoms, III, Nitrogen and Oxygen atoms', *The London, Edinburgh and Dublin Philosophical Magazine and Journal of Science*, 6th Series, Vol. 37, January–June 1919, pp. 571–80; E. Rutherford, 'On the Collision of α particles with Light Atoms, IV, An Anomalous Effect in Nitrogen', *The London, Edinburgh and Dublin Philosophical Magazine and Journal of Science*, 6th Series, Vol. 37, January–June 1919, pp. 581–87.
85 Discussed at length in Steven Krivit, 'The Rutherford Nitrogen-To-Oxygen Transmutation Myth', newenergytimes.com/v2/sr/Rutherford-Blackett/Rutherford-Blackett.shtml, accessed 29 January 2025.
86 E. Rutherford, 'Recent researches on the transmutation of the elements', *Nature*, 18 March 1933, pp. 388–89, esp. p. 388.
87 CUL RP Add 7653, Misc. files and notes, PA395-399, PA396 'Presidential Address', p. 20.
88 E. Rutherford, 'On the Collision of α particles with Light Atoms, IV, An Anomalous Effect in Nitrogen', esp. p. 586.
89 CUL RP Add 7653, Misc. files and notes, PA360, 'Lancashire Anti-Submarine Committee — Chairman's Dinner', 7 February 1919.
90 CUL RP Add 7653, Misc. files and notes, PA332-335, 'Reply to Admiralty, suggested draft for alteration or emendation', Rutherford's copy.

10 The crocodile

1 ATL MS-Papers 1342-258, Marsden, Ernest (Sir), 'Facts and particulars concerning the early life of Lord Rutherford', C.D. Ellis, 'The Cavendish Chair', typescript.
2 D. Schoenberg, 'Piotr Leonidovich Kapitza, 9 July 1894–8 April 1984', *Biographical Memoirs of Fellows of the Royal Society*, Vol. 31, Nov. 1985, pp. 326–74, p. 331.
3 Ibid.
4 CUL RP Add 7653, Correspondence L1-75, Langmuir to Rutherford, 10 December 1927.
5 CUL RP Add 7653, Correspondence P1-50, Pope to Rutherford, 11 March 1919.
6 ATL MS-Papers-1342-252, Marsden, Ernest (Sir) (1889–1970), Paper: Correspondence — Lord Rutherford, Rutherford to Schuster, 15 March 1919.
7 CUL RP Add 7653, Correspondence L1-75, Larmor to Rutherford, 9 March 1919.
8 Rutherford to his mother, 7 April 1919, quoted in Eve, p. 269.
9 CUL RP Add 7653, Correspondence N-O, Nagaoka to Rutherford 30 June 1919.
10 CUL RP Add 7653, Correspondence M238-264, Rutherford to Meyer 23 March 1919, with appended pencil drafts.
11 Ibid, 'Copy of resolution passed by the Senate, 8th May 1919.'
12 Ibid, Miers to Rutherford, 14 May 1919.
13 Campbell, pp. 379–80.
14 See e.g. ATL MS-Papers-3718, Rutherford, Ernest, 1st Baron Rutherford of Nelson 1871–1937: Letters to Professor A.W. Bickerton, Rutherford to Bickerton, 21 August 1925.
15 CUL RP Add 7653, Correspondence, Jeans to Rutherford, 12 September 1920. Einstein was confronted by substantial controversy in 1920 over his theory of General Relativity, including a major rally in August that filled the Berlin Philharmonic with opponents.
16 ATL Micro-MS-Coll-20-1939, Larmor, Lloyd, Lubbock and Rutherford Papers, Rutherford to Farr, 2 March 1936.
17 CUL RP Add 7653, Correspondence, L1-75, Larmor to Rutherford, 9 March 1919.
18 CUL RP Add 7653, Misc. files and notes, PA362, Rutherford's report on the history and the needs of the Cavendish laboratory, 'The needs of the Cavendish Laboratory'.
19 CUL RP Add 7653, Misc. files and notes, PA329, 'Hymn to Einstein' by 'AAR' (Alfred A. Robb).
20 CUL RP Add 7653, Misc. files and notes, PA301, Signed Menu for Cavendish Society Dinner, 10 December 1920.
21 CUL RP Add 7653, Correspondence Y-Z, Zeeman to Rutherford, 26 June 1920.
22 CUL RP Add 7653, Correspondence G1-80, Geiger to Rutherford, 26 March 1915.
23 Ibid, Geiger to Rutherford, 18 May 1919.
24 CUL RP Add 7653, Misc. files and notes, Misc. Xerox, Rutherford to Hevesey, 13 January 1920.
25 Ibid, Rutherford to Muller, 13 July 1919.
26 CUL RP Add 7653, Correspondence M161-233, Meyer to Rutherford, 22 January 1920.
27 CUL RP Add 7653, Correspondence E1-88, Ehrenhaft to Rutherford, 11 January 1924.
28 CUL RP Add 7653, Correspondence G1-80, Geiger to Rutherford, 19 November 1919.
29 ATL Micro-MS-Coll-20-1939, Larmor, Lloyd, Lubbock and Rutherford Papers, Rutherford to Marsden, 5 April 1922.
30 CUL RP Add 7653, Misc. files and notes, PA361, Lists of Projected Researches, 'Researches 1925–26', 'Projected Research August 1927', 'Projected researches 1927'.
31 E. Rutherford, 'Bakerian lecture: Nuclear Constitution of Atoms', *Proceedings of the Royal Society of London*, Series A, Vol. 97, No. 686, 1 July 1920, pp. 374–400, esp. p. 376.
32 D. Schoenberg, 'Piotr Leonidovich Kapitza, 9 July 1894–8 April 1984', *Biographical Memoirs of Fellows of*

the *Royal Society*, Vol. 31, Nov. 1985, pp. 326–74. For a summary of Kapitza's letters see also W. Boag, P.E. Rubinin, D. Schoenberg (eds), *Kapitza in Cambridge and Moscow: life and letters of a Russian physicist*, North Holland, Amsterdam, 1990.
33 Schoenberg, pp. 334–35.
34 CUL RP Add 7653, Correspondence K1-61, Rutherford to Kapitza, 9 October 1925 (typed).
35 Ibid, Kapitza to Rutherford, 3 May 1927.
36 https://mediatheque.lindau-nobel.org/laureates/kapitsa/research-profile, accessed 19 February 2025.
37 CUL RP Add 7653, Correspondence K1-61, Kapitza to Rutherford 17 December 1925.
38 CUL RP Add 7653, Misc. papers and notes PA363-374, E. Rutherford 'Report on the work of P. Kapitza, PhD', see also CUL RP Add 7653, Correspondence K1-61, Rutherford to Kapitza, 9 October 1925 (typed).
39 Malcolm Longair, 'Rutherford and the Cavendish Laboratory', Taylor & Francis Online, 1 November 2020, www.tandfonline.com/doi/full/10.1080/03036758.2021.1885452#d1e294, accessed 26 January 2025.
40 P.M.S. Blackett, 'The Rutherford Memorial Lecture, 1957', *Proceedings of the Royal Society*, London, Series A, No. 251, 9 June 1959, pp. 293–305, esp. p. 294.
41 P.M.S. Blackett, 'The Ejection of Protons from Nitrogen Nuclei, Photographed by the Wilson Method', *Proceedings of the Royal Society*, Series A, Vol. 107, No. 742, 2 February 1925, p. 357.
42 CUL RP Add7653, Rutherford Correspondence A1-A72, Andrade to Rutherford, 7 May 1937.
43 E. Rutherford, 'Bakerian lecture: Nuclear Constitution of Atoms', *Proceedings of the Royal Society of London*, Series A, Vol. 97, No. 686, 1 July 1920, pp. 374–400, esp. p. 375.
44 Ibid, p. 396.
45 Ibid, pp. 399–400.
46 CUL RP Add 7653, Correspondence M1-54, McCoy to Rutherford, 9 November 1920.
47 CUL RP Add 7653, Correspondence M161-233, Meyer to Rutherford, 6 October 1920.
48 Ernest Marsden, 'Rutherford at Manchester' in J.B. Birks (ed.), *Rutherford at Manchester*, Heywood & Company, London, 1962, p. 12.
49 CUL RP Add 7653, Misc. files and notes, PA322, 'Proton', handwritten note.
50 Sir E. Rutherford and J. Chadwick, 'The Disintegration of Elements by Particles', *The London, Edinburgh and Dublin Philosophical Magazine and Journal of Science*, 6th Series, Vol. 44, No. 261, September 1922, pp. 417–32, esp. pp. 421, 423 etc.
51 Chadwick, interview at web.archive.org/web/20141021094704/http://www.aip.org/history/exhibits/rutherford/sections/atop-physics-wave.html, accessed 1 February 2025.
52 Eve, p. 287.
53 Alexander Turnbull Library, ATL MS-Papers 1342-258, Marsden, Ernest (Sir), Facts and particulars concerning the early life of Lord Rutherford', C.D. Ellis, 'The Cavendish Chair', typescript.
54 CUL RP Add 7653, PA361, 'Researchers 1925–26', 'Projected researches 1927'.
55 P.M.S. Blackett, 'The Ejection of Protons from Nitrogen Nuclei, Photographed by the Wilson Method', *Proceedings of the Royal Society*, Series A, Vol. 107, No. 742, 2 February 1925, pp. 349–60, esp. pp. 356–59.
56 ATL Micro-MS-Coll-20-1939, Larmor, Lloyd, Lubbock and Rutherford Papers, Burbidge to Rutherford, 2 September 1925.
57 CUL RP Add 7653, Misc. files and notes, PA302, Pocket Notebook from Australian Tour 1925.
58 CUL RP Add 7653 Misc. files and notes, PA305a, Notes on W.S. Littlejohn (1934).
59 *Manawatu Standard*, 25 September 1925.
60 *Thames Star*, 28 September 1925.
61 See, e.g. CUL RP Add 7653, Rutherford Correspondence K1-61, Rutherford to Kapitsa, 9 October 1925.
62 *Te Aroha News*, 19 October 1925.
63 ATL Micro-MS-Coll-20-1939, Larmor, Lloyd, Lubbock and Rutherford Papers, Rutherford to Chadwick, 9 October 1925.
64 *Gisborne Times*, 27 October 1925.
65 *Press* (Christchurch), 31 October 1925.
66 *Evening Post*, 2 November 1925.
67 *New Zealand Herald*, 7 November 1925.
68 *Otago Daily Times*, 6 November 1925.
69 *New Zealand Times*, 12 November 1925.
70 ATL, MS-Papers-1342-255, Marsden, Ernest (Sir), 1889–1970: Papers, Correspondence re-Lord Rutherford, O.E. Rutherford to the Editor, *Atlantic*, 5 March 1959.
71 *Dominion*, 2 December 1925.
72 Summarised in Alain Aspect and Jacques Villain, 'La naissance de la mécanique ondulatoire', *Comptes Rendus Physique*, Vol. 18, Issues 9–10, November–December 2017, pp. 583–85.
73 W. Heisenberg, 'On the quantum reinterpretation of kinematical and mechanical relationships', trans. Luca Doria, www.staff.uni-mainz.de/doria/WH.pdf, accessed 28 January 2025.
74 See, e.g. CUL, RP PA 7653, Correspondence B417-465, de Broglie to Rutherford, 8 June 1922.
75 Observed by his student Charles Ellis in the 1920s, see ATL MS-Papers 1342-258, Marsden, Ernest (Sir), Facts and particulars concerning the early life of Lord Rutherford', C.D. Ellis, 'The Cavendish Chair', typescript.
76 CUL RP Add 7653, Notes and manuscripts, PA71 79, 'Abstract of Royal Institution Lecture, Friday March 19th 1932'.
77 André Siegfried, *America Comes of Age, a French Analysis*, trans. H.H. and Doris Hemmong, Harcourt, Brace and Co., New York, 1927. This translation was published in April 1927 and by December 1928 had gone through 13 printings.
78 CUL RP Add 7653, Correspondence L1-75, Langmuir to Rutherford, 10 December 1927.
79 M. Stanley Livingston, 'The history of the cyclotron', Proceedings of the 7th International Conference on Cyclotrons and their Applications, Zurich, Switzerland, accelconf.web.cern.ch/c75/papers/j-01.pdf, accessed 31 January 2024.
80 John D. Cockcroft, 'Experiments on the interaction of high-speed nucleons with atomic nuclei', Nobel Lecture, 11 December 1951, www.nobelprize.org/uploads/2018/06/cockcroft-lecture.pdf, accessed 1 February 2025.
81 John Douglas Cockcroft and E.T.S. Walton, 'Experiments with high velocity positive ions', *Proceedings of the Royal Society*, Series A, 3 November 1930, p. 480.
82 http://waywiser.fas.harvard.edu/objects/14566/kenotron-rectifier, accessed 1 February 2025.

83 John Douglas Cockcroft and E.T.S. Walton, 'Experiments with high velocity positive ions', p. 481.
84 James Chadwick, 'The Neutron and its Properties', Nobel Lecture, 12 December 1935, www.nobelprize.org/uploads/2018/06/chadwick-lecture.pdf, accessed 1 February 2025.
85 John D. Cockcroft, 'Experiments on the interaction of high-speed nucleons with atomic nuclei', Nobel lecture, December 11, 1951, www.nobelprize.org/uploads/2018/06/cockcroft-lecture.pdf, accessed 1 February 2025
86 CUL RP Add 7653, Correspondence M55-101a, Rutherford to Marsden 12 October 1932.
87 Ibid.
88 Paul Adrien Maurice Dirac, 'Quantised singularities in the electromagnetic field', *Proceedings of the Royal Society*, Series A, Vol. 133, Issue 821, September 1931, pp. 60–72, esp. pp 61–62.
89 For details see Carl D. Anderson, 'The production and properties of positrons', Nobel Lecture, 12 December 1936, www.nobelprize.org/uploads/2018/06/anderson-lecture.pdf.
90 E. Rutherford, 'Recent researches on the transmutation of the elements', *Nature*, 18 March 1933, pp. 388–89, esp. p. 389.
91 ATL Micro-MS-Coll-20-1939, Larmor, Lloyd, Lubbock and Rutherford Papers, Rutherford to Marsden, 3 November 1934.

11 Elder statesman

1 Quoted in Eve, p. 342.
2 See, e.g. CUL RP Add 7653, Correspondence R75-137, Jim Rutherford to Rutherford, 3 January 1909, "Dear Ern ..."
3 ATL MS-Papers-1342-251, Marsden, Ernest (Sir), 1889–1970: Papers, Original correspondence relating to Rutherford, Bay Rutherford to Martha Rutherford, 19 January 1931, 2 February 1930, etc.
4 Ibid, Bay Rutherford to Martha Rutherford, 17 February 1931.
5 CUL RP Add 7653, Misc. files and notes, PA318-331, 'Congratulations answers.'
6 ATL Micro-MS-Coll-20-1939, Larmor, Lloyd, Lubbock and Rutherford Papers, Bledisloe to Rutherford, 11 April 1931; see also CUL RP Add 7653, Correspondence B54-96, Bledisloe to Rutherford, 11 April 1931.
7 CUL RP Add 7653, Correspondence M238-264, Mules to Rutherford, 1 January 1931. Underline and double quote marks in original.
8 CUL RP Add 7653, Correspondence T1-54, Thompson to Rutherford, 1 January 1931.
9 CUL RP Add 7653, PA318, 'Congratulations answers'.
10 Eve, p. 341.
11 CUL RP Add 7653, Correspondence, B417-465, Broad to Rutherford, 15 February 1931.
12 CUL RP Add 7653, Correspondence M1-54, McLennan to Rutherford, 2 February 1931.
13 CUL RP Add 7653, Correspondence G1-80, Geiger to Rutherford, 23 January 1931.
14 CUL RP Add 7653, Correspondence B297-353, Born to Rutherford, September 1931.
15 CUL RP Add 7653, Correspondence B1-B53, Baldwin to Rutherford, 16 October 1932; see also associated letter sequence.
16 CUL RP Add 7653, Correspondence A1-72, Walter Adams to Rutherford, 29 April 1934.
17 See, e.g. CUL RP Add 7653, Correspondence D1-8,1 Donnan to Rutherford, 19 November 1935.
18 CUL RP Add 7653, Rutherford Correspondence M1-54, Rutherford to McGowan (ICI), 6 February 1936.
19 CUL RP Add 7653, Misc. files and notes PA332-335, 'The wandering scholars'.
20 CUL RP Add 7653, Correspondence, D1-81, Rutherford to Donnan, 6 June 1936.
21 CUL RP Add 7653, Correspondence, K1-61, Kapitza to Rutherford, 4 September 1933.
22 See, e.g. ATL MS-Papers-1342-252, Marsden, Ernest (Sir) 1889–1970, Papers, Correspondence — Lord Rutherford, Kapitza to Rutherford, 19 October 1935.
23 CUL RP Add 7653, Correspondence U1, Rutherford to Urey, 20 June 1933; see also CUL RP ADD 7653, 5 Lot 03, T. Knox-Shaw to M. Y. Sokoloff, 23 November 1936, and other correspondence in file.
24 CUL RP Add 7653, Correspondence, G1-80, Gamow to Rutherford, 5 March 1934.
25 ATL MS-Papers-1342-252, Marsden, Ernest (Sir) 1889–1970, Papers, Correspondence — Lord Rutherford, Rutherford to Grey, 10 May 1934.
26 CUL RP Add 7653, Correspondence N-O, Oliphant to Rutherford, 7 January 1934.
27 CUL RP Add 7653, Rutherford Correspondence, W1-81, Walker to Rutherford, 10 November 1934.
28 The first flights to Australia were initiated as early as 1919. A weekly service from Croydon to the major eastern Australian cities was initiated in April 1935 by Imperial Airways and Qantas Empire Airways.
29 *Auckland Star*, 17 July 1935; obituary, *Nelson Evening Mail*, 17 July 1935.
30 CUL RP Add 7653, Misc. files and notes, PA307, Membership Card of Pontifica Academi Scientorum.
31 CUL RP Add 7653, Correspondence M55-101a, Miela to Rutherford with enclosure, 22 March 1937.
32 ATL Micro-MS-Coll-20-1939, Larmor, Lloyd, Lubbock and Rutherford Papers, Rutherford to Marsden, 8 October 1934.
33 CUL RP Add 7653, Correspondence G1-80, Rutherford to Geiger, 5 September 1937.
34 Print copy in CUL RP Add 7653, Misc. files and notes, PA 395-399, 'Presidential Address', p. 20. This was ultimately read to the congress by James Jeans.
35 Eve, p. 424.
36 ATL MS-Papers-1342-256, Marsden, Ernest (Sir), 1889–1970, Papers, Correspondence re-Lord Rutherford, James Rutherford to Marsden, 13 November 1937.
37 CUL RP Add 7653, Correspondence N-O, Oliphant to Rutherford, 11 October 1937. Whether Rutherford saw this is unclear: the letter was typed, dated 11 October. Post from Birmingham to Cambridge was typically one day, so Rutherford likely received it on 13 or 14 October.
38 N. Feather, 'Rutherford Memorial Lecture 1977: some episodes of the α-particle story', *Proceedings of the Royal Society*, London, Series A, Vol. 357, pp. 117–29, p. 117.
39 ATL MS-Papers-1342-256, Marsden, Ernest (Sir), 1889–1970, Papers, Correspondence re Lord Rutherford, James Rutherford to Marsden, 13 November 1937.
40 Ibid.

41 Eve, pp. 425–26.
42 Sir Mark Oliphant, reading notes by Chadwick, 'Some personal recollections of Rutherford, The Man', *Notes and Records of the Royal Society of London*, Vol. 27, Issue 1, August 1972, pp. 7–23, p. 8.
43 Eve, p. 426.
44 'Bicentenary of the birth of Galvani', *Nature*, 13 November 1937.
45 Campbell, p. 474.
46 *New Zealand Herald*, 22 October 1937.
47 CUL RP Add 7653, Misc. files and notes, PA406, Mackintosh to Cockcroft, 21 October 1937.
48 Ibid, Holden to Cockcroft, 20 October 1937.
49 Noted in Campbell, p. 373.
50 CUL RP Add 7653, Misc. files and notes PA309, funeral service.
51 *Press*, 23 October 1937.
52 Eve, p. 428.
53 ATL MS-Papers-1342-256, Marsden, Ernest (Sir), 1889–1970, Papers, Correspondence re-Lord Rutherford, Mary Rutherford to James Rutherford, 5 December 1937.

12 Ernest Rutherford and the birth of modern physics

1 E. Marsden, 'Rutherford — His Life and work, 1871–1937', Rutherford Memorial Lecture, 1954, *Proceedings of the Royal Society of London*. Series A. Mathematical and Physical Sciences, Vol. 226, Issue 1166, November 1954, p. 284.
2 ATL MS-Papers-1342-256, Marsden, Ernest (Sir), 1889–1970, Papers, Correspondence re-Lord Rutherford, James Rutherford to Marsden, 22 October 1937.
3 Ibid, Marsden to Mary Rutherford, 28 October 1937.
4 Ibid, Marsden to Mary Rutherford, 4 November 1937.
5 Ibid, Marsden to Farr, 8 November 1937.
6 ATL MS-Papers- [illegible] to Marsden, 3 October 1961; Peter du Sautoy to Marsden, 5 December 1958.
7 Notably MS-Papers-1342-258, Marsden, Ernest (Sir), 1889–1970: Papers, Facts and Particulars concerning the early life of Lord Rutherford and MS-Papers-1342-259, Marsden, Ernest (Sir), 1889–1970: Papers, Facts and Particulars concerning the early life of Lord Rutherford.
8 Quoted in Eve, p. 4.
9 ATL MS-Papers-1242-258, Marsden, Ernest (Sir), 1889–1970: Papers, 'Facts and particulars concerning the early life of Lord Rutherford', typescript, C.C. Farr, 'Early Years in New Zealand', notes by G. Gillespie.
10 Quoted in Eve, p. 10.
11 Noted in Robert P. Murray, 'The 1896 Magnetic Detector of Lord Rutherford', www.antiquewireless.org/homepage/, accessed 3 October 2024.
12 Noted in Eve, p. 268.
13 ATL MS-Papers 1342-258, Marsden, Ernest (Sir), Facts and particulars concerning the early life of Lord Rutherford', C.C. Farr, 'Famous New Zealanders — Lord Rutherford', typescript.
14 Ernest Marsden, 'Rutherford at Manchester' in Birks (ed.), p. 13.

Glossary

Alchemy The idea held across Medieval Europe that matter could be transformed, typically into valuable materials such as gold.

Alpha decay The process by which a parent nucleus disintegrates by emitting an alpha particle, a helium nucleus. This particle is short-range and is strongly ionising, because each particle rapidly gains up to two electrons from particle encounters, rendering the target atom an ion. After acquiring these electrons an alpha particle becomes a helium atom. This phenomenon also limits the range and penetrating power of alpha particles. It is also why much of the world's helium is found near major deposits of uranium or thorium ores.

Antimatter One prediction of quantum mechanics was that matter should have a mirror-image. The first anti-particle was discovered in 1932, proving the mathematics. One of the continuing questions in modern physics is why the universe doesn't have equal quantities of matter and antimatter.

Becquerel The Standard International (SI) unit for a quantity of radiation, defined as one decay event per second.

Beta radiation Rutherford discovered this in 1897, calling his discovery 'beta rays' to distinguish them from the 'alpha rays' he found at the same time. He found them to be 100 times more penetrating than alpha rays. In 1899, Henri Becquerel used J.J. Thomson's electrical methodology to discover that beta rays were high-speed electrons. Today we know that they are an outcome of weak-force interation (qv) and are produced when a neutron inside an unstable atomic nucleus collapses into a proton, an electron, and a neutrino anti-particle. They have medical application in cancer treatment.

Classical physics The mechanisms defined by Isaac Newton and his colleagues to describe the workings of the world they observed, including gravity. This was superseded from the late ninteenth century.

Crookes tube The name given to a device able to generate X-rays, devised by Sir William Crookes and widely used by 1890s physicists. The principle was later developed first into the oscilloscope tube, then TV picture tubes.

Curie A pre-Standard International definition for a specific quantity of radiation, first devised in 1910 by Rutherford and colleagues and named after Pierre and Marie Curie. It has since been redefined as 3.7×10^{10} decay events per second.

Cyclotron A type of particle accelerator that uses a spiral track.

Decay chain The sequence of isotopes (qv) and elements (qv) that emerge as radioactive material breaks down by emitting particles from its nucleus. Each has a specific half-life (qv).

Electrodynamics The study of electromagnetic forces in motion. This became a major focus of study for physicists in the late nineteenth century, who soon saw it as a fundamental way of understanding all that they were observing. It was superseded as an 'organising system' for physics by quantum mechanics.

Electromagnetism One of the fundamental forces of the universe, defined in the mid-nineteenth century by James Maxwell and forming the basis of the physics revolution that followed. Most modern technology involves the principles of electromagnetism, one way or another.

Electron A subatomic particle discovered by J.J. Thomson in 1897, the first subatomic particle whose existence was confirmed.

Electron volt A measure of kinetic energy – the energy of a moving object – defined as the energy gained by an electron moving through an electric potential difference of one volt in vacuum. It is a very tiny figure, equal to 1.602×10^{-19} joules.

Emanation A term used by Rutherford during his initial researches to describe unknown emissions from radioactive elements, without bias as to their nature.

Euclidean geometry The geometry of the everyday world around us in which distances and angles are both fixed and measurable. Named after the classical Greek philosopher Euclid.

Fundamental forces A generic term used to describe the basic forces underpinning everything observed across the universe. Three have been proven: electromagnetism, strong and weak force (qv). A fourth force, gravity, is often included although Einstein's theory of General Relativity makes clear it is a second-order effect, not a force of itself.

Gamma radiation Gamma radiation is the third and most powerful form of radiation emitted by radioactive materials and is the highest energy available in the electromagnetic spectrum. Gamma radiation is emitted by some atomic nuclei, but it can also be generated in other ways.

General relativity This is the explanation developed by Albert Einstein that explains space, time, gravity and their relationships, and from that the workings of the entire observable macro-scale universe. It has been proven correct by every test, including through phenomena such as gravity waves that Einstein predicted, but which were not observed until the twenty-first century. Although repeatedly shown to be correct, general relativity does not play nicely with the accepted explanation for quantum effects.

Half-life A term coined by Rutherford to measure the rate of radioactive breakdown. Specifically, it refers to the rate at which emissions fall to half an original level. This continues indefinitely: 1/2, 1/4, 1/8, 1/16, 1/32, 1/64, 1/128 and so forth.

Ion An atom or molecule which has either lost or gained an electron, thus giving it a positive or negative electric charge. Hence *ionising*: causing ions to form by forcing an extra electron or electrons upon an atom or by taking electron/s away. Radiation powerful enough to knock electrons loose is known as ionising radiation.

Isotope A term devised by Frederick Soddy to describe different versions of the same element. Subsequent discoveries, led by Rutherford, showed that these were a result of differing numbers of neutrons in the nucleus, giving isotopes different atomic weights.

LHC The Large Hadron Collider operated near Geneva by Conseil Européen pour la Recherche Nucléaire (CERN, the European Organisation for Nuclear Research). At time of writing it is the world's most powerful synchotron (qv) and obtains data in a way basically identical to Rutherford's experimental apparatus of the 1910s, smashing particles together in order to observe what emerges. The energies involved, however, are much higher than Rutherford could obtain with what he had available.

Neutron An electrically neutral particle of the hadron family, found within an atomic nucleus and predicted by Rutherford in 1920. The following year a US chemist suggested such a neutral particle could be called a 'neutron'. At Rutherford's suggestion his colleague James Chadwick conducted experiments in 1932 that confirmed the existence of these particles. All atoms except some isotopes of hydrogen contain at least one neutron.

Nucleus Discovered by Rutherford in 1911, a nucleus holds the bulk of the mass of any atom and comprises a mixture of protons and neutrons.

Planck constant A constant, h, identified by Max Planck that remains the basic value in quantum mechanics. For example, the energy in a photon is equal to the frequency multiplied by the Planck constant: $E = hf$.

Positron A positively charged electron, see *antimatter*.

Proton A positively charged particle within an atomic nucleus, discovered by Rutherford during experimentation late in the First World War, following up on work he had instructed his associates to conduct. Named by him after William Prout. It has 1836 times the mass of an electron.

Radioactivity A term coined by Marie Curie to describe the emissions observed from heavy elements such as uranium, originally as a hyphenated word 'radio-activity'.

Radiometric dating A technique in which Rutherford was a key pioneer, whereby the products of radioactive breakdown can be used to specifically date objects that contain this material. Rutherford was the first to provide an age for the Earth that was of the same general magnitude – billions of years – as modern measurements. This principle is also used in radiocarbon dating.

Quantum mechanics The system progressively devised from 1900 by Max Planck and Albert Einstein, and then post-First World War by a range of physicists including Erwin Schrödinger, Werner Heisenberg and Niels Bohr, to explain what was being observed in terms of energy steps – hence the word 'quantum'. The word 'mechanics' was used instead of 'physics' by its earliest advocates because what they saw was so different from classical physics. This system has since been extended to become the fundamental explanation for everything observed by physicists, except gravity and time. Quantum effects have been shown to be real, however the explanation adopted in the late 1920s, the 'Copenhagen' interpretation, does not correlate with Einstein's General Relativity, which has also been shown to be correct. This remains to be resolved at time of writing.

Special relativity This theory was devised by Albert Einstein in 1905, working from electrodynamic principles proposed by George FitzGerald, to explain how things worked if the speed of light was constant at all times.

Standard Model The Standard Model is based on quantum field theory and defines all known particles and their properties. This is the closest physicists have come to a 'theory of everything', although it still does not describe every observed phenomenon. It includes 17 fundamental particles that between them define mass and carry energy. No particle has been found to carry either gravity or time, and the interactions described by the Standard Model do not cover gravity.

Strong force This is the strongest of the fundamental forces, variously measured but considered to be around 137 times stronger than electromagnetism and 10^{38} times stronger than gravity. It operates primarily at distances within the nucleus. This force is primarily responsible for the energies released by radioactive decay, which are enormous by comparison with chemical reactions.

Synchotron A type of particle accelerator in which particles are accelerated around a circular track, enabling very high energies to be achieved. The LHC (qv) is of this type. The energy lost by forcing the particles around a curved path is known as synchotron radiation.

Wave-function The term devised during the 1920s to define the quantum state of a particle.

Weak force One of the fundamental physical forces of the universe, not known to Rutherford but now understood to be part of the process of radioactive decay. It operates at distances less than the diameter of a proton and allows fundamental particles to transform, for example, during beta decay when a neutron emits an electron and an electron anti-neutrino, thus becoming a proton. Weak force also plays a role in nuclear fusion in stellar cores; and it is responsible for some forms of radioactive transmutation via a process known as electron capture, in which an electron falls into the nucleus and enables a proton to transform into a neutron, facilitated by the weak force.

Bibliography

Primary sources

NATIONAL LIBRARY OF NEW ZEALAND

Alexander Turnbull Library (ATL)

Micro MS-Coll-20-2016, Personal collection of Sir William Bragg (continued).

Micro-MS-Coll-20-1939, Larmor, Lloyd, Lubbock and Rutherford papers.

Micro-MS-Coll-20-2015, Records of Secretary: Personal collections of Sir Humphry Davy, Michael Faraday, Sir William Grove, John Tyndall and Sir William Bragg.

MS-Papers-10930-2, Rutherford family: papers, George Rutherford diaries.

MS-Papers-1342-251, Marsden, Ernest (Sir), 1889–1970: Papers, Original correspondence relating to Rutherford.

MS-Papers-1342-252, Marsden, Ernest (Sir), 1889–1970, Papers, Correspondence — Lord Rutherford.

MS-Papers-1342-254, Marsden, Ernest (Sir), 1889–1970: Papers, Correspondence between Rutherford and Marsden.

MS-Papers-1342-256, Marsden, Ernest (Sir), 1889–1970: Papers, Correspondence re-Lord Rutherford.

MS-Papers-1342-258, Marsden, Ernest (Sir), 1889–1970: Papers, Facts and Particulars concerning the early life of Lord Rutherford.

MS-Papers-1342-259, Marsden, Ernest (Sir), 1889–1970: Papers, Facts and Particulars concerning the early life of Lord Rutherford.

MS-Papers-1342-260, Marsden, Ernest (Sir), 1889–1970, Papers, Addresses and articles on Lord Rutherford.

MS-Papers-1342-262, Marsden, Ernest (Sir), 1889–1970: Papers, Addresses and articles written by E. Marsden re-Lord Rutherford.

MS-Papers-1342-263, Marsden, Ernest (Sir), 1889–1970: Papers, Addresses and Articles by E. Marsden re-Lord Rutherford.

MS-Papers-3718, Rutherford, Ernest, 1st Baron Rutherford of Nelson 1871–1937: Letters to Professor A.W. Bickerton.

CAMBRIDGE UNIVERSITY LIBRARY (CUL)

Rutherford Papers Add 7653 (by folio)

Correspondence

A1-72
B1-53
B54-96
B97-169
B170-229
B417-465
C1-53
C54-115
D1-81
E1-88
F1-66
G1-80
H1-88
H89-125
I-J
K1-61
L1-75
L76-146
M1-54
M55-101a
M102-160
M161-233
M238-264
N-O
P1-50
R1-74
R75-137

S1-88
T1-54
T55-96
U1
W1-81
Y-Z

Drafts of published papers

Miscellaneous files and notes

PA294-299
PA301
PA302
PA304-305a
PA306
PA307
PA308
PA309
PA312
PA314-315
PA318-331
PA332-335
PA361
PA362
PA363-374
PA375
PA381-394
PA395-399
PA406

Laboratory notebooks

NB1-6
NB7-16
NB17-25

Notes and manuscripts

PA6A-6B
PA4A-5
PA35-37
PA194 (part)
Misc. Xerox

Rutherford Papers Add 7654 (by folio)

Correspondence R.

LIBRARY AND ARCHIVES CANADA

Passenger lists 1865–1922.

PUBLISHED PRIMARY SOURCES

Badash, Lawrence (ed.), *Rutherford and Boltwood: letters on radioactivity*, Yale University Press, New Haven, 1969.

Marder, Arthur (ed.), *Fear God and Dread Nought: the Correspondence of Admiral of the Fleet Lord Fisher of Kilverstone*, Vol. III, Jonathan Cape, London, 1959.

SECONDARY SOURCES

Academic papers and theses

Barham, L., Duller, G.A.T., Candy, I. et al., 'Evidence for the earliest structural use of wood at least 476,000 years ago', *Nature*, 622, 107–11 (2023). doi.org/10.1038/s41586-023-06557-9.

Blackett, P.M.S., 'The Ejection of Protons from Nitrogen Nuclei, Photographed by the Wilson Method', *Proceedings of the Royal Society*, Series A, Vol. 107, No. 742, 2 February 1925.

Bohr, N., 'On the Constitution of Atoms and Molecules', *The London, Edinburgh and Dublin Philosophical Magazine and Journal of Science*, 6th Series, Vol. 26, July–December 1913.

Brown, Samantha, et al., 'The earliest Denisovans and their cultural adaptation', *Nature*, 25 November 2021, www.nature.com/articles/s41559-021-01581-2.

Cockcroft, John Douglas and E.T.S. Walton, 'Experiments with high velocity positive ions', *Proceedings of the Royal Society*, Series A, 3 November 1930.

Degano, Ilaria, Sylvain Soriano, Paola Villa, Luca Pollarolo, Jeannette J. Lucejko, Zenobia Jacobs, Katerina Douka, Silvana Vitagliano, Carlo Tozzi, 'Hafting of Middle Paleolithic tools in Latium (central Italy): New data from Fossellone and Sant'Agostino caves', doi.org/10.1371/journal.pone.0213473.

Dewar, James, 'The rate of production of helium from radium', *Proceedings of the Royal Society of London*, Series A, Vol. 81, Issue 507, September 1908.

Draper, John Draper, 'On the production of light by heat', *The London, Edinburgh and Dublin Philosophical Magazine and Journal of Science*, Series 3, Vol. 30, July–December 1847.

Einstein, A., '1st die Tragheit eines Korpers von seinem Energiegehalt abhangig?', *Annalen der Physik*, 17, 1905.

—, 'Concerning the Investigation of the State of Aether in Magnetic Fields', c.1894, 1library.net/document/y9jgg4vq-concerning-the-investigation-of-the-state-of-aether-in-magnetic-fields-by-albert-einstein.html.

—, 'On a heuristic point of view concerning the production and transformation of light', English translation at einsteinpapers.press.princeton.edu/vol2-trans/100.

—, 'Considerations concerning the fundaments of theoretical physics', *Science*, Vol. 91, No. 2369, May 1940.

Geiger, H. and E. Marsden, 'On a diffuse reflection of the α-particles', *Proceedings of the Royal Society of London*, Series A, Vol. 82, No. 557, July 1909.

Heisenberg, W., 'On the quantum reinterpretation of kinematical and mechanical relationships', trans. Luca Doria, www.staff.uni-mainz.de/doria/WH.pdf.

Hublin, Jean-Jacques, Simon Neubauer, and Philipp Gunz, 'Brain ontogeny and life history in Pleistocene hominins ', *Philosophical Transactions of the Royal Society, London*, B. Biol. Sci., 5 March 2015; 370(1663): 20140062 at www.ncbi.nlm.nih.gov/pmc/articles/PMC4305163/.

Jeans, James, 'The Mechanism of Radiation', *The London, Edinburgh and Dublin Philosophical Magazine and Journal of Science*, 6th Series, Vol. 2, Jul–Dec 1901.

Kozowyk, P.R.B., Baron, L.I. and Langejans, G.H.J., 'Identifying Palaeolithic birch tar production techniques: challenges from an experimental biomolecular approach', *Sci Rep 13*, 14727 (2023), doi.org/10.1038/s41598-023-41898-5.

Langley, Michelle C. and Thomas Suddendorf, 'Archaeological evidence for thinking about possibilities in hominin evolution', doi.org/10.1098/rstb.2021.0350.

Maclaurin, J.S., 'On the Action of Potassium-Cyanide Solution upon Gold', *Transactions and Proceedings of the Royal Society of New Zealand*, Vol. 28, 1895.

Marquet J-C., Freiesleben T.H., Thomsen K.J., Murray A.S., Calligaro M., Macaire J-J. et al., (2023), 'The earliest unambiguous Neanderthal engravings on cave walls: La Roche-Cotard, Loire Valley, France'. *PLoS ONE* 18(6): e0286568.

Maxwell, J. Clerk, FRS, [James Maxwell], 'A dynamical theory of the electromagnetic field', *Philosophical Transactions of the Royal Society of London*, Vol. 155, 1865, pp. 459–512

Maxwell, James, 'On Faraday's Lines of Force', *Transactions of the Cambridge Philosophical Society*, Vol. X, Part I.

Michelson, Albert A. and Edward W. Morley, 'On the Relative Motion of the Earth and the Luminiferous Ether', *The American Journal of Science*, Vol. XXXIV, No. 203, November 1881.

Moseley, H.G.J., 'The High-Frequency Spectra of the Elements', *The London, Edinburgh and Dublin Philosophical Magazine and Journal of Science*, 6th Series Vol. 26, July–December 1913.

Murray, Robert P., 'The 1896 Magnetic Detector of Lord Rutherford', www.antiquewireless.org/homepage/.

Owens, R.B., 'Thorium radiation', *The London, Edinburgh and Dublin Philosophical Magazine and Journal of Science*, Vol. 48, July–December 1899.

Pickering, William H., 'A Geiger counter study of the cosmic radiation', PhD thesis, California Institute of Technology, Pasadena, California, 1936.

Planck, Max, 'Ueber das Gesetz der Energieverteilung im Normalspectrum', *Annalen der Physik*, Vol. 309, No. 3, 1901.

Rutherford, E., 'Einfluss der Temperatur auf die "Emanationen" Radioaktiver Substanzen', *Physikalische Zeitschrift*, No. 29, 20 April 1901.

—, 'On the Collision of α particles with Light Atoms, I, Hydrogen', *The London, Edinburgh and Dublin Philosophical Magazine and Journal of Science*, 6th Series, Vol. 37, January–June 1919.

—, 'On the Electrification of gases exposed to Röntgen Rays, and the Absorption of Röntgen Radiation by gases and vapours', *The London, Edinburgh and Dublin Philosophical Magazine and Journal of Science*, 5th Series, Vol. 43, January–June 1897.

—, 'Radioactivity produced in Substances by the Action of Thorium Compounds', *The London, Edinburgh and Dublin Philosophical Magazine and Journal of Science*, 5th Series, Vol. 49, January–June 1900.

—, 'Some Remarks on Radioactivity', *The London, Edinburgh and Dublin Philosophical Magazine and Journal of Science*, 6th Series No. 5, January–June 1903.

—, 'A Radio-active Substance emitted from Thorium Compounds', *The London, Edinburgh and Dublin Philosophical Magazine and Journal of Science*, 5th Series, Vol. 49, January–June 1900.

—, 'Bakerian Lecture: Nuclear Constitution of Atoms', *Proceedings of the Royal Society of London*, Series A, Vol. 97, No. 686, 1 July 1920.

—, 'Magnetization of Iron by High-frequency Discharges', *Transactions and Proceedings of the New Zealand Institute*, Vol. XXVII, Wellington, May 1895.

—, 'On the Collision of α particles with Light Atoms, II, Velocity of the hydrogen atom', *The London, Edinburgh and Dublin Philosophical Magazine and Journal of Science*, 6th Series, Vol. 37, January–June 1919.

—, 'On the Collision of α particles with Light Atoms, III, Nitrogen and Oxygen atoms', *The London, Edinburgh and Dublin Philosophical Magazine and Journal of Science*, 6th Series, Vol. 37, January–June 1919.

—, *The Collected Papers of Lord Rutherford of Nelson, Volume 2: Manchester*, Routledge, New York, 2016 (paperback).

—, 'The structure of the atom', *The London, Edinburgh and Dublin Philosophical Magazine and Journal of Science*, 6th Series, Vol. 27, January–June 1914.

—, 'Uranium Radiation and the Electrical Conduction produced by it', *The London, Edinburgh and Dublin Philosophical Magazine and Journal of Science*, 5th Series, Vol. 47, January–June 1899.

—, MA. BSc, 1851 Exhibition Science Scholar, 'Magnetic Viscosity', *Transactions and Proceedings of the Royal Society of New Zealand*, Vol. 28, 1895.

—, 'On the Collision of α particles with Light Atoms, IV, An Anomalous Effect in Nitrogen', *The London, Edinburgh and Dublin Philosophical Magazine and Journal of Science*, 6th Series, Vol. 37, January–June 1919.

Rutherford, E. and H. T. Barnes, 'Heating Effect of the Radium Emanation', *The London, Edinburgh and Dublin Philosophical Magazine and Journal of Science*, 6th Series, January–June 1904.

—, and Miss H.T. Brooks, 'The New Gas from Radium', *Proceedings and Transactions of the Royal Society of Canada*, Series 7, Vol. 7, Section III, 1901.

Rutherford, Sir E. and J. Chadwick, 'The Disintegration of Elements by Particles', *The London, Edinburgh and Dublin Philosophical Magazine and Journal of Science*, 6th Series, Vol. 44, No. 261, September 1922.

—, and H. Geiger, 'An electrical method of counting the number of α-particles from radio-active substances', *Proceedings of the Royal Society of London*, Series A, Vol. 81, Issue 546, August 1908.

—, and H. Geiger, 'The charge and nature of the α-particle', *Proceedings of the Royal Society of London*, Series A, Vol. 81, Issue 546, August 1908.

—, and F. Soddy, 'On Radioactive Change', *The London, Edinburgh and Dublin Philosophical Magazine and Journal of Science*, 6th Series, No. 5, January–June 1903.

—, and Soddy, F., 'The Cause and Nature of Radioactivity, Pt 1', *The London, Edinburgh and Dublin Philosophical Magazine and Journal of Science*, 6th Series, Vol. 4, July–December 1902.

Schmidt, Patrick, Radu Iovita, Armelle Charrié-Duhaut, Gunther Möller, Abay Namen, Ewa Dutkiewicz, 'Ochre-based compound adhesives at the Mousterian type-site document complex cognition and high investment', *Science Advances*, 21 February 2024, Vol. 10, Issue 8, DOI: 10.1126/sciadv.adl082.

Siefert, Antonia, Zachary G. Lane, Marco Galoppo, Ryan Ridden-harper, David L. Wiltshire, 'Supernovae evidence for foundational change to cosmological models', *Monthly Notices of the Royal Astronomical Society: Letters*, Vol. 537, Issue 1, February 2025.

Shah, Jayant, 'Accuracy of Ptolemy's Almagest in predicting solar eclipses', *Annals of Mathematical Science and Applications*, Vol. 3, No. 1, 2018.

Stokes, G.G., 'On the Constitution of the Luminiferous Aether, viewed with reference to the phenomenon of the Abberation of Light', *The London, Edinburgh and Dublin Philosophical Magazine and Journal of Science*, Vol. 29, July–December 1846.

Stokes, G.G., 'On the Constitution of the Luminiferous Aether', *The London, Edinburgh and Dublin Philosophical Magazine and Journal of Science*, Vol. 32, January–June 1848.

Thomson, J.J. 'Cathode Rays', *The London, Edinburgh and Dublin Philosophical Magazine and Journal of Science*, 5th Series, Vol. 44, 1897.

—, 'On the Structure of the atom', *The London, Edinburgh and Dublin Philosophical Magazine and Journal of Science*, Vol. 7, No. 49, January–June 1904.

Thomson, J.J. and E. Rutherford, 'On the Passage of Electricity through Gases exposed to Röntgen Rays', *The London, Edinburgh and Dublin Philosophical Magazine and Journal of Science*, 5th Series, Vol. 42, July–December 1896.

Tuniz, C., F. Bernadini, I. Turk, L. Dimkaroski, L. Mancini, D. Dreossi, 'Did Neanderthals play music? X-ray computed micro-tomography of the Divje babe "flute"', doi.org/10.1111/j.1475-4754.2011.00630.x.

NEWSPAPERS

Auckland Star
Colonist
Dominion
Evening Star

Fielding Star
Gisborne Times
Manawatu Times
Marlborough Press
Nelson Evening Mail
New York Times
New Zealand Times
Opunake Times
Otago Witness
Patea Mail
Press (Christchurch)
Star (Christchurch)
Sun (Christchurch)
Taranaki Herald
Thames Star
The Times

BOOKS AND PERIODICALS

— 'Bicentenary of the birth of Galvani', *Nature*, 13 November 1937.

— 'Scientists debate planet definition and agree to disagree', www.psi.edu/blog/scientists-debate-planet-definition-and-agree-to-disagree/.

— *La structure de la Matière, Rapports et discussions du Conseil de Physique*, Gauthier-Villars, Paris, 1921.

— *The Physical Laboratories of Manchester: a record of 25 years work*, Manchester University Press, Manchester, 1906.

— *La Structure de la Matière, rapports et discussions du Conseil de physique tenu à Bruxelles du 27 au 31 Octobre 1913, sous les auspices de l'Institut international de physique Solvay*, Gauthier-Villars et Co., Paris, 1921.

Anderson, Carl D., 'The production and properties of positrons', Nobel Lecture, 12 December 1936, www.nobelprize.org/uploads/2018/06/anderson-lecture.pdf.

Andrade, E.N. da C., *Rutherford and the Nature of the Atom*, Anchor Books, New York, 1964.

Aspect, Alain and Jacques Villain, 'La naissance de la mécanique ondulatoire', *Comptes Rendus Physique*, Vol. 18, Issues 9–10, November–December 2017.

Bacon, Francis, 'Of Studies', *The Essays Or Counsels, Civil And Moral, Of Francis Ld Verulam Viscount St. Albans*, via Project Gutenberg, www.gutenberg.org/files/575/575-h/575-h.htm.

Badash, Lawrence, 'Influence of New Zealand on Rutherford's scientific development', in Nathan Reingold and Marc Rothenberg (eds), *Scientific Colonialism: a cross-cultural comparison*, Smithsonian Institution Press, Washington, 1987.

Bailly, Charles, 'Atomic modelling in the early 20th century: 1904–1913', University of Colorado, Boulder, Department of Physics, 12 October 2008, http://philsci-archive.pitt.edu/4232/1/Atomic_Modeling_101208.pdf.

Baldwin, Melinda, 'Ernest Rutherford's Ambitions', *Physics Today*, Vol. 75, No. 5, May 2021.

Berends, Fritz and Franklin Lambert, 'Einstein's Witches', *Europhysics News*, July 2018, paperity.org/p/100465520/einsteins-witches.

Birks, J.B. (ed.), *Rutherford at Manchester*, Heywood & Company, London, 1962.

Blackett, P.M.S., 'The Rutherford Memorial Lecture, 1957', *Proceedings of the Royal Society of London*, Series A, No. 251, 9 June 1959.

Bohr, Niels, 'The Solvay Meetings and the Development of Quantum Mechanics', Niels Bohr at the occasion of the 12th Solvay Conference in physics, 9–14 October 1961, http://www.solvayinstitutes.be/pdf/Niels_Bohr.pdf.

Born, Max, *Einstein's Theory of Relativity*, revised edition, Dover Publications, New York, 1962.

Brooke, Geoffrey, 'Real Wages in New Zealand: 1840–1914', www.nzae.org.nz/wp-content/uploads/2011/08/Real_Wages_in_New_Zealand_1840-1914.pdf, accessed 17 September 2023.

Brown, Eric N. and Dan. L. Borovina, 'The Trinity High-Explosive Implosion system: the Foundation for Precision Explosive Applications', *Nuclear Technology*, Vol. 207, Supplement 1, 2021.

Campbell, John, *Rutherford: scientist supreme*, AAS Publications, Christchurch, 1999.

Chadwick, James, 'Some personal recollections of Rutherford, The Man', *Notes and Records of the Royal Society of London*, Vol. 27, Issue 1, August 1972.

—, 'The Neutron and its Properties', Nobel Lecture, 12 December 1935, www.nobelprize.org/uploads/2018/06/chadwick-lecture.pdf.

Churchill, Winston, *My Early Life*, Thornton Butterworth, London, 1930.

Cockcroft, John D., 'Experiments on the interaction of high-speed nucleons with atomic nuclei', Nobel Lecture, 11 December 1951, www.nobelprize.org/uploads/2018/06/cockcroft-lecture.pdf.

D'Agostino, Salvo, 'Hertz's Researches on Electromagnetic Waves', *Historical Studies in the Physical Sciences*, Vol. 6, 1975.

Davis, Brian R., 'Maclaurin, James Scott', *Dictionary of New Zealand Biography*, teara.govt.nz/en/biographies/3m23/maclaurin-james-scott.

Devons, S., FRS, 'Rutherford and the science of his day', *Notes and Records of the Royal Society of London*, Vol. 45, No. 2, (1991).

Devrees, Jozeph T., 'The light came in 1905', arxiv.org/pdf/physics/0602083.

Dirac, Paul Adrien Maurice, 'Quantised singularities in the electromagnetic field', *Proceedings of the Royal Society*, Series A, Vol. 133, Issue 821, September 1931.

Eve, A.S., *Rutherford, being the life and letters of the Rt Hon Lord Rutherford, OM*, Cambridge University Press, Cambridge, 1939.

Fairbank, Jeremy C.T., 'William Adams and the spine of Gideon Algernon Mantell', *Annals of the Royal College of Surgeons of England*, 2004, Vol. 686.

Feather, N., 'Rutherford Memorial Lecture 1977: some episodes of the α-particle story', *Proceedings of the Royal Society*, London, Series A, Vol. 357.

Fisher, Admiral Sir John (Lord Fisher), *Records*, Hodder & Stoughton, London, 1919.

Fowler, Gilbert J., 'The Silver Jubilee of the Indian Science Congress', *Current Science*, Vol. 6, No. 7, January 1938.

Fronde, Clifford and Robert Meyrowitz, 'Studies on uranium minerals: Rutherfordine, Diderchite and Clareite', *Trace Elements Investigations Report 474*, United States Department of the Interior Geological Survey, October 1954.

Gershon, Livia, 'The X-ray craze of 1896', daily.jstor.org/the-x-ray-craze-of-1896/.

Gribbin, John, *Einstein's Master-Work: 1915 and the General Theory of Relativity*, Icon Books, London, 2015.

Haubold, Hans, 'Albert A. Michelson's Experientum Crucis 1881 in Potsdam, Germany', arxiv.org/pdf/2111.12176.

Heilbron, J.L., *Ernest Rutherford and the explosion of atoms*, Oxford University Press, Oxford, 2003.

Hughes, J., 'Rutherford, Radioactivity and the origins of nuclear physics', *Journal of Physics*, conference series, No. 381, 2012.

Katzir, Shaul, 'Who knew piezoelectricity? Rutherford and Langevin on submarine detection and the invention of sonar', *Notes and Records of the Royal Society*, Vol. 66, 2012.

Kitcher, Philip, 'Fluxions, Limits and Infinite Littleness: a study of Newton's presentation of the calculus', *Isis*, Vol. 64, No. 1, March 1973.

Klien, Martin J., 'Max Planck and the beginnings of quantum theory', *Archive for the History of Exact Sciences*, Vol. 1, No. 5, 1962.

Kragh, Helge, 'Rutherford, Radioactivity and the atomic nucleus', Centre for Science Studies, Aarhus University, Aarhus, Denmark (n.d.), arxiv.org/pdf/1202.0954.

Krivitt, Steven, 'The Rutherford Nitrogen-To-Oxygen Transmutation Myth', newenergytimes.com/v2/sr/Rutherford-Blackett/Rutherford-Blackett.shtml.

Kuhn, Thomas E., *The Structure of Scientific Revolutions*, University of Chicago Press, Chicago, 1962, third edition 1996.

Langevin, P. and M. de Broglie, *La Theorie du Rayonnement et les Quanta, rapports et discussions*, Guthier-Villars, Paris, 1912.

Langevin-Jolot, H., 'Radium, Marie Curie and Modern Science', *Radiation Research*, No. 150 (Supplement), S3-S8, 1998.

Livingston, M. Stanley, 'The history of the cyclotron', Proceedings of the 7th International Conference on Cyclotrons and their Applications, Zurich, Switzerland (n.d.).

Lloyd Prichard, M.F., *An Economic History of New Zealand to 1939*, Collins, Auckland, 1970.

Longair, Malcolm, 'Rutherford and the Cavendish Laboratory', *Journal of the Royal Society of New Zealand*, 51(3–4), pp.444–66, at doi.org/10.1080/03036758.2021.1885452.

Longino, Helen, *The Fate of Knowledge*, Princeton University Press, Princeton, 2002.

Mackay, Ruddock, *Fisher of Kilverstone*, Oxford University Press, Oxford, 1973.

MacLeod Roy M. and E. Kay Andrews, 'Scientific advice in the War at Sea, 1915–1917: the Board of Invention and Research', *Journal of Contemporary History*, Vol. 6, No. 2, 1971.

Malley, Marjorie, 'The discovery of atomic transmutation: scientific styles and philosophies in France and Britain', *Isis*, Volume 70, No. 2, June 1979.

Marsden, Ernest, 'Rutherford — His Life and Work, 1871–1937', Rutherford Memorial Lecture, 1954, *Proceedings of the Royal Society of London*. Series A, Vol. 226, Issue 1166, November 1954.

—, 'Rutherford at Manchester' in J.B. Birks (ed.), *Rutherford at Manchester*, Heywood & Company, London, 1962.

Marshall, James L. and Virginia R. Marshall, 'Ernest Rutherford: the "true discoverer" of Radon, *Bulletin for the History of Chemistry*, Vol. 28, No. 2, 2003.

—, 'Rutherford and radon', *The Hexagon*, Summer 2010.

Martins, Roberto de Andrade, *Historical Studies on Radioactivity*, Quamcumque Editum, 2021.

Mavani, Himanshu and Navinder Singh, 'A concise history of the black-body radiation problem', arxiv.org/pdf/2208.06470, April 2022.

Museum of Radiation and Radioactivity, 'How the Curie Came to Be', orau.org/health-physics-museum/articles/how-the-curie-came-to-be.html.

Niaz, Mansoor, 'Evolving Nature of Objectivity in the History of Science and its Implications for Science Education', 2018; 46: 37–77, DOI: 10.1007/978-3-319-67726-2_3.

Pattison, Michael, 'Scientists, Inventors and the Military in Britain 1915–19: the Munitions Inventions Department', *Social Studies of Science*, Vol. 13, 1981.

Perrett, W. and G.B. Jeffery (trans.), *The Principle of Relativity*, Dover Publications, reprinting Methuen & Co., 1923.

Radvanyi, Pierre, 'The discussion between P. Curie and E. Rutherford (1900–1904), *The European Physical Journal*, Vol. 38, 2013.

Reeves, Richard, *A Force of Nature: the frontier genius of Ernest Rutherford*, Norton, New York, 2008.

Robinson, H.R., 'Rutherford: Life and work to the year 1919, with personal reminiscences of the Manchester period' in J.B. Birks (ed.), *Rutherford at Manchester*, Heywood & Company, London, 1962.

Rutherford, E., 'Henry Gwyn Jeffreys Moseley', *Nature*, 9 September 1915.

—, 'Recent researches on the transmutation of the elements', *Nature*, 18 March 1933.

—, 'The constitution of matter and the evolution of the elements', *Popular Science Monthly*, Vol. 87, August 1915.

—, 'The recent radium controversy', Letters, *Nature*, 25 October 1906.

—, 'Recent researches on the transmutation of the elements', *Nature*, 18 March 1933.

—, 'Emanations from radio-active substances', *Nature*, 13 June 1901.

—, 'The international radium standard', *Nature*, 4 April 1912.

—, 'International conference on the structure of matter', *Nature*, 20 November 1913.

—, *Radioactivity*, Cambridge University Press, Cambridge, 1904.

—, ed. J.A. Ratcliffe, 'The development of the theory of atomic structure', in Joseph Needham and Walter Pagel (eds), *Background to Modern Science*, Cambridge University Press, Cambridge, 1938.

—, *Radioactive Transformations*, Yale University Press, New Haven, 1906.

—, and H.T. Barnes, 'Heating Effect of the Radium Emanation', Letters, *Nature*, 29 October 1903.

Saul, John Ralston, *Voltaire's Bastards: the dictatorship of reason in the west*, Penguin, London, 1993.

Schoenberg, D., 'Piotr Leonidovich Kapitza, 9 July 1894–8 April 1984', *Biographical Memoirs of Fellows of the Royal Society*, Vol. 31, November 1985.

Shapin, Steven, *The Scientific Revolution*, University of Chicago Press, Chicago, 1996.

Siegfried, André, *America Comes of Age, a French Analysis*, trans. H.H and Doris Hemmong, Harcourt, Brace and Co., New York, 1927.

Soddy, Frederick, 'Reminiscences of McGill, 1900–1902', *Old McGill*, Vol. 36, McGill University, Montreal, 1933.

Stamenkovic, Phillipe, 'Facts and objectivity in science', *Interdisciplinary Science Reviews*, 48(2), pp. 277–98. doi.org/10.1080/03080188.2022.2150807.

Stewart, Balfour, *Physics*, D. Appleton & Company, New York, 1878.

Straussman, Norbert, 'On Einstein's Doctoral Thesis', www.researchgate.net/publication/2172866_On_Einstein's_Doctoral_Thesis.

Süsskind, Charles, 'Hertz and the technological significance of electromagnetic waves', *Isis*, Vol. 53, No. 6, Autumn 1965.

—, 'Observations of Electromagnetic Wave Radiation before Hertz', *Isis*, Vol. 55, No. 1, March 1964.

Terroux, F.R., 'The Rutherford Collection of Apparatus at McGill University', *Transactions of the Royal Society of Canada*, S. III, 1938.

Todd, Neil, 'Historical and radio-archaeological perspectives on the use of radioactive substances by Ernest Rutherford', University of Manchester, 1 December 2008.

Travis, Anthony S., 'The accidental discovery of mauve', *Victorian Review*, Vol. 40, No. 2, Johns Hopkins University Press, 2014.

W.L.C., 'Geology at the British Association', *Nature*, 7 October 1915.

Weinstein, Galina, 'Einstein, Schwarzchild, the Perehelion Motion of Mercury and the Rotating Disk story', Tel Aviv University, 25 November 2014.

Weizmann, Chaim, *Trial and Error: the autobiography of Chaim Weizmann*, Vol. 1, Jewish Publication Society of America, Philadelphia, 1949.

Wells, H.G., 'The World Set Free', www.gutenberg.org/cache/epub/1059/.

—, *The War of the Worlds*, Harper & Brothers, New York, 1898.

Witteveen, Joeri, 'Objectivity, Historicity, Taxonomy', *Erkenntnis*, Vol. 83, pp. 445–63, 2018, via doi.org/10.1007/s10670-017-9897-z.

Wright, Matthew, *Guns and Utu*, Penguin, Auckland, 2011.

—, *Old South: life and times in the nineteenth century mainland*, Penguin, Auckland, 2009.

—, *The New Zealand Experience in Gallipoli and the Western Front*, Oratia Books, Auckland, 2017.

—, *Those Who Have the Courage*, Oratia Books, Auckland, 2024.

—, *Waitangi: A Living Treaty*, Bateman Books, Auckland, 2018.

—, *Western Front: the New Zealand Division 1916–1918*, Reed, Auckland, 2005.

—, *Illustrated History of New Zealand*, 3rd edition, Bateman Books, Auckland, 2023.

—, *The History of Hawke's Bay*, Intruder Books, Wellington, 3rd edition 2024

Websites

aps.org
bankofengland.co.uk
bdmhistoricalrecords.dia.govt.nz
blog.sciencemuseum.org.uk/jj-thomsons-cathode-ray-tube
community.netweather.tv
daily.jstor.org
edition.cnn.com
einsteinpapers.press.princeton.edu
energyeducation.ca/encyclopedia
france-inflation.com
gamelab.mit.edu
havelock.school.nz
history.aip.org
historyofinformation.com
interestingengineering.com
makingscience.royalsociety.org
mbie.govt.nz
mediatheque.lindau-nobel.org
mhs.ox.ac.uk
ncbi.nlm.nih.gov/pmc/articles/PMC3520298
nobelprize.org
numericana.com
nzas.co.nz
nzedge.com
nzrugby.co.nz
orau.org/health-physics-museum
phy.cam.ac.uk
physics.stackexchange.com
royalsociety.org
solvay.com
spark.iop.org/rutherford
ssmaritime.com
teara.govt.nz
thecanadianencyclopedia.ca
theprow.org.nz
waywiser.fas.harvard.edu
web.archive.org
www.hyperphysics.phy-astr.gsu.edu/hbase

Index

K

L

M

N

Ø

O